# Algal Diversity at Jayakwadi Bird Sanctuary

***The Author***

**Professor Dr. Prakash B. Papdiwal** has retired from the Botany Department at Dr Babasaheb Ambedkar Marathwada University,Chhatrapati Sambhajinagar, Maharashtra. With extensive teaching and research experience, he has made significant contributions to algae, plant bacteriology, seed pathology and tree pathology. He has published three books, over 100 research papers in journals of repute, and guided 13 Ph D students. A fellow of Indian Phytopathological Society and Indian Botanical Society, Dr Papdiwal was on the editorial board of eight research journals. He was President of Indian Phytopathological Society's (Western Zone).

# Algal Diversity at Jayakwadi Bird Sanctuary

*– Author –*

**Prakash B. Papdiwal**

*Former Professor, Department of Botany*
*Dr. Babasaheb Ambedkar Marathwada University*
*Chhatrapati Sambhaji Nagar – 431004*
*Maharashtra*

**2026**

**Daya Publishing House®**
*A Division of*
**Astral International Pvt. Ltd.**
**New Delhi – 110 002**

ISBN: 978-93-5919-277-2



*Published by* : **Daya Publishing House®**
*A Division of*
**Astral International Pvt. Ltd.**
– ISO 9001:2015 Certified Company –
4736/23, Ansari Road, Darya Ganj
New Delhi-110 002
Ph. 011-43549197
e-mail: info@astralint.com
Website: www.astralint.com

*Printed at* : **Replika Press Pvt Ltd**

# Acknowledgement

The author is thankful to University Grants Commission, New Delhi for the financial assistance under Major Research Project "Studies on plant diversity of Nathsagar" (F. No. 39-431/2010(SR) dated 06-01-2011). The present text is the part of work carried out on algal diversity at Nathsagar (Jayakwadi Bird Sanctuary).

The author is thankful to the Professor and Head, the Department of Botany, Dr. Babasaheb Ambedkar Marathwada University, Chhatrapati Sambhaji Nagar for providing laboratory facilities.

I am thankful to the Officers and Staff Members of Wildlife (WL) Division, Chhatrapati Sambhaji Nagar, Govt. of Maharashtra; for their help in various ways. I am thankful to Mr. Samir Sahai, the then Conservator of Forests (working plan), Chhatrapati Sambhaji Nagar for taking initiative in the study of biodiversity at Nathsagar. The permission given by him for collection of samples and their study is gratefully acknowledged. I also wish to put on record the help rendered by Mr. Mukul Trivedi, the then Deputy Conservator of Forests (WL), Chhatrapati Sambhaji Nagar; for the present study. It is also essential to thank Mr. S. R. Kulkarni, Retd. Assistant Conservator of Forests [Jayakwadi Bird Sanctuary (J.B.S.)] and Mr. P.W. Deshpande, the then Assistant Conservator of Forests (J.B.S.) for their help in present study.

I appreciate the work carried out by Dr. S. B. Andhale and Dr. P. B. Jadhawar, who have carried out Ph. D. research on algal diversity at Jayakwadi Bird Sanctuary, under my guidance. The help given by Miss. S. G. Bandgar, Project Fellow, in the preparation of the project report is acknowledged with thanks. Some of the illustrations are from the thesis of Andhale (2008).

I am thankful to Prof. Digambar Rao, Kakatiya University, Warangal and Prof. Rashmi Singh, Govt. Autonomous P.G. College, Satna (M.P.) for critically reviewing this text.

The efforts taken by Mr. Soham S. Badjate in the preparation of this manuscript is appreciated.

***Prakash B. Papdiwal***

# Introduction

Jayakwadi project is constructed on Godavari River at Paithan, about 50 kms away from Chhatrapati Sambhaji Nagar. Geographically it is located between 75°00′00″ and 75° 15′00″ E longitudes and 90°18′13″ and 19°33′16″ N latitudes. The river basin is very wide and the terrain is very flat. Therefore, constructed wall of dam is about 10280 meters in length. The total height of the dam is about 37.73 meters. The gross storage of water is 2,909 M.Cu.M. and submerged area of the water reservoir is 339.80 sq. km, at its maximum capacity. There are 30 islands of various sizes across the shallow waters. The dam receives water jointly by the rivers Godavari and Pravara. The weather in the project area is dry. The relative humidity is low fluctuating between 25 and 30 per cent, except during the monsoon. During rainy season the relative humidity may reach upto 80 per cent. Maximum temperature during summer is around 42°C while minimum 6°C during winter, whereas, the average temperature is about 25°C. The monsoon breaks up in the middle of June and continuous up to the end of September. On an average, the annual rainfall varies from 616 to 880 mm. The winter season is short, which starts in November and ends up in early February.

The water reservoir of the Jayakwadi dam is very large, which is about 55 kms in length and 27 kms in width, which is called as Nathsagar. Except the main stream of the river, the water is shallow in the dam. The sunlight reaches up to the bottom of the dam in most of the area (25 to 28 per cent), and therefore, the biomass productivity of water reservoir is high. The water body has good number of algae, pteridophytes angiospermic plants, fishes and other aquatic animals. Every year, during winter, nearly 250 different types of migratory birds visit the dam. Taking this into consideration, the water body was declared as a Wildlife Sanctuary under the Wildlife Protection Act of 1972 by the Govt. of Maharashtra. About 33,980 hectares of the reservoir and 125 hectares of Gyaneshwar Udyan became part of the Jayakwadi bird sanctuary on 10$^{th}$ October, 1986- [Deputy Conservator of Forests (Wildlife), Chhatrapati Sambhaji Nagar].

It appears that due to favorable environmental factors, ample food and safety, this habitat is at the peak in fetching the birds from November to February every year. These birds have different food habits. They show varying degree of adaptation towards their food habits. Birds like ducks are herbivorous, which use the aquatic vegetation as their food (Kamat, 1967; Salim Ali, 2002; Saini, 2005; Rajalakshmi *et al.*, 2007). Some birds like Brown-capped pygmy woodpecker are omnivorous, which feed on the aquatic flora as well as fauna. Majority of the migratory birds are carnivorous, which use the aquatic animals, especially fishes, as their food. The fishes can also be classified as herbivorous, carnivorous and omnivorous. The herbivorous fishes use aquatic vegetation, especially algae as their food. The carnivorous fishes use herbivorous and omnivorous fishes as their food. The omnivorous fishes survive on the aquatic vegetation as well as on small fishes. Therefore, the aquatic vegetation including algae is the primary producer in the water body. The fishes and migratory birds are dependent on aquatic vegetation for their food.

There are about 457 National parks and wildlife sanctuaries in India, of these 46 are the bird sanctuaries. The Nathsagar water reservoir, is one among them with large number of algae and aquatic plants which serve as food for fishes and the migrating birds.

A search through the available literature revealed that very little work has been carried out on the aquatic flora of freshwater bodies, which have been declared as bird sanctuaries. Therefore, the topic has been selected for present investigation. This is the first report of algal flora of Nathsagar.

# Algal Diversity at Nathsagar

### Class: Chlorophyceae

### Order: Chlorococcales

### Family: Chlorococcaceae

### Genus: *Chlorococcum* Meneghini

**1) *Chlorococcum humicola* (Nageli) Rabenhorst (Pl. 1, Fig. 1)**

Philipose, 1967, p 73, f 3

Cell spherical, solitary or a number of cells crowded together to form a stratum. Chloroplast like a hollow sphere with a lateral notch. Cells 15 μ in diameter.

**2) *C. infusionum* (Schrank) Meneghini (Pl. 1, Fig. 2)**

Philipose, 1967, p 73, f 1

Cells usually spherical, rarely ovoid or elongated and of variable dimensions. Chloroplast like a hollow sphere with a notch on one side. Cells 85 μ in diameter.

### Family: Hydrodictyaceae

### Genus: *Hydrodictyon* Roth

**3) *Hydrodictyon indicum* Iyengar (Pl. 1, Fig. 3)**

Philipose, 1967, p 134, f 49

Cells meshes are much larger; cells are very turgid and forming nets, therefore, very brittle; cells deeply green, cylindrical and very big; cell wall very thick and lamellated with knob like portions of the cell wall.

### 4) *H. reticulatum* (Linn.) Lagerheim (Pl. 1, Fig. 4)

Philipose, 1967, p 134, f 48

Colonies macroscopic, free floating, single layered, net like, reticulate, meshes pentagonal or hexagonal. Cells elongated, cylindrical, Cells up to 800 μ long and up to 150 μ broad. Nets up to 20 cm long.

## Genus: *Sorastrum*

### 5) *Sorastrum americanum* (Bohlin) Schmidle v. *undulatum* G.M. Smith (Pl. 1, Fig. 5)

Prescott, 1951, p 228, pl 50, f 10

Cells with outer free walls emarginate and furnished at each of 4 angles with a long stout, outwardly directed spine; cells attached to centre of the colony. Cells 15 μ in diameter, 12.5 μ long, and spines 10 μ long.

### 6) *S. spinulosum* Nageli (Pl. 1, Fig. 6)

Philipose, (1967), p 132-133, f 47

Colonies spherical, free-floating of 8 reniform cells, three angled and with a short stalk from the basal angle and two short pointed spines from each of the outer angles. Chloroplast parietal with a single pyrenoid. Cells 8-20μ broad, 6-18 μ long. Spines 4-8μ long. Eight-celled colony 26.4 μ in diameter.

## Genus: *Pediastrum*

### 7) *Pediastrum angulosum* (Her.) Meneghini v. *laevigatum* Raciborski (Pl. 2, Fig. 1)

Philipose, 1967, p 118, f 39

Colonies 8-16-32 celled, flat, compact, without perforations. Interior cells irregularly hexagonal. Marginal cells transversely elongated, wider above, truncated at the base; outer side deeply emarginated,with slightly converging lobes with obtuse or rounded ends. Colonies up to 150μ in diameter.

### 8) *P. boryanum* (Turpin) Menegh. (Pl. 2, Fig. 2)

Philipose, 1967, p 118, f 40 (a)

Colonies circular to oval and 16 celled, cells arranged in concentric rings without intercellular spaces. Inner cells polygonal with straight sides. Outer face of marginal cells slightly to deeply emarginate and with two short processes ending in stumpy spines. Cell wall usually granulate, sixteen celled colonies 80 μ in diameter.

### 9) *P. boryanum* v. *brevicorne* Braun. (Pl. 2, Fig. 3a, b, c and d)

Patel and I. George, 1977, p 173, f 1

Colonies flat, more or less circular or oval, 4-8-16 or 32 celled, compact and without perforation. Outer cells with two short processes. Inner cells polygonal, with straight margins. Cell wall smooth or granulate. Cells 25.2μ broad and 23.4μ long. Colony up to 108.4μ broad.

### 10) *P. boryanum* v. *longicorne* Reinsch (Pl. 2, Fig. 4a, b)

Philipose 1967, p119-120, f 40b

Colonies flat, circular to oval and usually of 8-16 cells arranged in concentric rings without intercellular spaces. Cells with longer processes than in the type. Inner cells polygonal with straight sides. Outer face of marginal cells slightly to deeply emarginated with two processes. Cells 10.2-14μ in diameter. Colonies up to 102 μ in diameter.

### 11) *P. duplex* Meyen (Pl. 2, Fig. 5)

Philipose, 1967, p 121, f 43 (a-b)

Colony 16 celled, cells with small lens shaped perforations between cells. Inner cells quadrate to angular and not in contact at the central portion of the side cells. Inner side of marginal cells concave, outer side produced into two short truncate processes. Cells 15 μ in diameter, colony up to 45 μ in diameter.

### 12) *P. duplex* Meyen v. *clathratum* (A. Braun) Lagerheim (Pl. 2, Fig. 6)

Philipose, 1967, p 123, f 43 (e)

Cells with more deeply emarginated sides and larger intercellular spaces. Colony 16 celled. Cells 12.5 μ in diameter, colony 60 μ in diameter.

### 13) *P. duplex* Meyen v. *genuinum* (A. Braun) Hansgirg (Pl. 2, Fig. 7)

Philipose, 1967, p 123, f 43 (d)

Colony 32 celled with fairly large intercellular spaces. Marginal cells with stout processes which are straight. Cell membrane smooth. Cells 10 μ, perforations 5 μ and colony 62.5 μ in diameter.

### 14) *P. duplex* Meyen v. *reticulatum* Lagerheim (Pl. 2, Fig. 8)

Philipose, 1967, p 124, f 43 (g)

Cell more or less H-shaped with sides of processes of marginal cells nearly parallel. Intercellular spaces large and oval. Cells 17.5 μ in diameter; 16 celled colony 65 μ in diameter.

### 15) *P. duplex* Meyan v. *gracillimum* W.et. G.S. West (Pl. 2, Fig. 9)

Philipose 1967, p 124-125, f 43 h and I

Colonies flat, with very large intercellular spaces. Cells very narrow, as broad as or narrower than the processes. Body of marginal cells curved outwards and with two long processes with emarginated apices. Inner cells also similar to marginal cells but with shorter processes. Cells 15-25μ in diameter, perforations 4-16μ in diameter. Colonies 65-140 μ in diameter.

### 16) *P. duplex* Meyen v. *subgranulatum* Raciborski (Pl. 3, Fig. 1)

Philipose, 1967, p 125, f 43 (j)

Colony 16 celled. Colony with intercellular spaces.Cell wall distinctly granulate. Cells 10 μ in diameter, colony 50 μ in diameter.

### 17) *P. glanduliferum* Bennett. (Pl. 3, Fig. 2a and b)

Prescott 1951, p 224, f. 4

Colonies flat, circular, not perforated. entire or nearly so; peripheral cells 5 or 6 angled with a concave incision between 2 long processes, capitate projection tipped with a globular swelling; cells 10.8-11.5μ in diameter,

### 18) *P. integrum* Naegeli (Pl. 3, Fig. 3)

Prescott, 1951, p 225, pl 48, f 10

Colony entire; cells 5 sided; outer margin of peripheral cells smooth or with two short and much reduced processes, and granular walls, emarginate between the processes; cells 17.5 μ in diameter.

### 19) *P. ovatum* (Ehr.) A.Braun (Pl. 3, Fig. 4a and b)

Philipose, 1967, p 115, f 37

Colonies usually 16 celled, with the cells arranged in a ring round a central space or with one or more interior cells and number of marginal cells, perforate or almost imperforate, perforation being small. Cell wall smooth or ornamented.

### 20) *P. sculptatum* G.M. Smith. (Pl. 3, Fig. 5)

Prescott, 1951, p 226, pl 49, f 5

Colony entire or narrow perforations; internal cells 4-6 sided; peripheral cells with two lobes having subparallel margins; cell wall covered with a fine reticulum of ridges; cell 10 μ in diameter; colony 157.5 μ wide.

### 21) *P. simplex* Meyen (Pl. 3, Fig. 6)

Philipose, 1967, p 113, f 36

Colonies circular to oval, 4 to 32 celled. Inner side of marginal cells nearly straight, outer side produced into a gradually tapering process, sides concave.

Inner cell polygonal. Cells in contact with adjacent ones and usually without inter cellular spaces. When present, intercellular spaces very small and few in number. Cell wall smooth or punctate to granulate. Cells 12.5 µ broad, 22.5 µ long.

### 22) *P. simplex* Meyen v. *duodenarium* (Bailey) Rabenhorst (Pl. 3, Fig. 7)

Philipose, 1967, p 115, f 36 (d-h)

Colony flat, 16 celled. Differs from the type in having large intercellular spaces or single central space with the cells arranged in a ring at the pheriphery. Inner face of marginal cells concave, outer face prolonged into single delicately tapering processes. Sides of marginal cells also concave or nearly straight. Interior cells similar to marginal cells but with shorter processes. Cell wall smooth or finely punctate. Cells 20 µ in diameter, colonies 112.5 µ in diameter.

### 23) *P. simplex* (new variety) (Pl. 3, Fig. 8)

Colonies flat, circular, 16 celled with large intercellular spaces. Outer cells drawn into a single tapering process. It is distinct from the typical variety *simplex* in having the processes curved at the terminal ends. Cells 18-25.2 µ in diameter. Colony 70 µ in diameter.

### 24) *P. tetras* (Ehr.) Ralfs. (Pl. 3, Fig. 9)

Prasad and Misra, 1992, p 11, pl 1, f 9

Colonies circular, 8 celled, lacking intercellular spaces and perforation on wall; outerside of marginal cells with a deep, cuneate incision upto the middle of cell; each lobe slightly truncate, inner cell fixed with a single linear incision. Colony 25 µ; long. cell 7.5 µ, and lat. cell 10 µ. in diameter

### 25) *P. tetras* (Her.) Ralfs. v. *excisum* (Rabenh.) Hansgirg (Pl. 3, Fig. 10)

Philipose, 1967, p 129, f 45 (f)

Colony rectangular, 4 celled; lobes are deeply concave. Cells 10 µ broad and long, colony 20 µ broad and long.

### 26) *P. tetras* (Ehr.) Ralfs. v. *tetraodon* (Corda) Hansgirg (Pl. 3, Fig. 11)

Philipose, 1967, p 129, f 45 (g)

Colonies 16 celled. Incision of cells deep with the lobe adjacent to the incision of marginal cells very pronounced. Cells 15 µ in diameter.

## Sub Family: Tetraedronoideae

## Genus: *Tetraedron* Kuetzing

### 27) *Tetraedron caudatum* (Corda) Hansgirg (Pl. 4, Fig. 1)

Philipose, 1967, p 150, f 64 (b)

Cells small, flat, five sided with four of the sides concave and the fifth in the form of notch of varying depth. Angles rounded and produced into a small straight spine. Cells 15 μ in diameter, spine 2.5 μ long.

### 28) *T. minimum* (A.Braun) Hansgirg (Pl. 4, Fig. 2)

Prescott, 1951, p 267, pl 60, f 12

Cells small, flat, tetragonal, the angles rounded and without spine or processes, lobes sometimes cruciately arranged; margins of the cell concave, with one frequently incised; cells 15 μ in diameter.

### 29) *T. muticum* (A.Braun) Hansgirg f. *punctulatum* (Reinsch) (Pl. 4, Fig. 3)

Prescott, 1951, p 268, pl 60, f 18

Cells flattened, triangular, the angles sharp but without spines; margins slightly convex; wall granular; cells 15 μ in diameter, 7.5 μ thick.

### 30) *T. regulare* Kuetzing v. *bifurcatum* Wille (Pl. 4, Fig. 4)

Prescott, 1951, p 269, pl 61, f 1

Cells tetragonal, pyramidal, the sides convex or slightly concave, the angles broadly rounded and tipped with two stout often curved spines; cells 35 μ in diameter including spines.

### 31) *T. regulare* Kuetzing v. *granulata* Prescott (Pl. 4, Fig. 5)

Prescott, 1951, p 269, pl 61, f 3

Cells large, pyramidal, the lobes broad and stout with convex margins, the angles broadly rounded and tipped with a single short spine which may be reduced to a mere papilla; wall punctuate and covered with small granules or roughening; cells 37.5 μ in diameter.

### 32) *T. trigonum* (Naegali) Hansg. (Pl. 4, Fig. 6)

Prasad and Misra, 1992, p16, pl 1, f 13

Cells flat, triangular with slightly concave sides, angles rounded, each ending in a stout spine; cell wall without any ornamentations; cell 30 μ; long spines 5 μ.

## Family : Oocystaceae

## Sub Family: Oocystoideae

## Genus: *Oocystis* Naegeli

### 33) *Oocystis borgei* Snow (Pl. 5, Fig. 1)

Philipose, 1967, p 183, f 93

Cells broadly ellipsoid with rounded ends. Poles not thickened. Chloroplast 1-4, parietal, each with a pyrenoid. Cells 12.5 μ broad, 15 μ long, usually in 2-8 celled colonies, the enclosing envelope being more or less round and narrow.

### 34) *O. crassa* Wittrock (Pl. 5, Fig. 2)

Philipose, 1967, p 181, f 91 (b)

Cells solitary or in colonies of 2-4, ellipsoidal, nearly twice as long as broad and with mammillary thickenings at the poles. Chloroplast parietal, fairly large, each cell with a pyrenoid. Cell 20 μ broad and 25 μ long.

### 35) *O. ecballocystiformis* Iyengar (Pl. 5, Fig. 3)

Philipose, 1967, p 186, f 99 (a)

Cells oblong, ellipsoid with broadly rounded ends. Cell membrane thick and without polar thickenings. Chloroplast parietal and disc shaped, each with a minute pyrenoid. Cells 22.5 μ broad, 15 μ long.

### 36) *O. elliptica* W. West (Pl. 5, Fig. 4)

Philipose, 1967, p 186, f 100 (b)

Colony 4 celled with envelope narrow, rarely solitary. Cells elongate-ellipsoid with broadly rounded ends which are not thickened. Chloroplast numerous and in the form of parietal discs without pyrenoid. Cells 12.5 μ broad, 20 μ long.

### 37) *O. gigas* Archer (Pl. 5, Fig. 5)

Prasad and Misra, 1992, p 20, pl 2, f 3

Colonies of 4 cells; cells broadly ellipsoid with rounded ends poles not thickened; chloroplast parietal, discoid, pyrenoid not seen. Long cell. 25 μ; lat cell 15 μ.

### 38) *O. irregularis* (Petkof) Printz (Pl. 5, Fig. 6)

Philipose, 1967, p 184, f 95

Cells irregularly ellipsoid to round and usually crowded towards one side. Cell wall thin and without apical thickenings. Chloroplast single, parietal, covering the cell completely and without a pyrenoid. Cells 17.5 μ broad, 25 μ long.

### 39) *O. naegelii* A. Braun (Pl. 5, Fig. 7)

Prasad and Misra, 1992, p 20, pl 2, f 11

Colonies 4 celled with closely fitting envelope; cell ovoid to oblong, without polar nodules; cell membrane thick; Chloroplast one in form of parietal plate, lacking pyrenoid. Long cell 22.5 µ; lat cell 15 µ.

### 40) *O. pyriformis* Prescott (Pl. 5, Fig. 8)

Prescott, 1951, p 246, pl 54, f 8

Cells broadly pyriform-ovoid, with a prominent apiculation at one pole, the other end broadly rounded; united in families of two or four; chloroplast massive and parietal with one pyrenoid; cells 15 µ in diameter, 17.5 µ long; colony 35 µ in diameter, 47.5 µ long.

### 41) *O. solitaria* Wittrock (Pl. 5, Fig. 9)

Philipose, 1967, p 180, f 89 (a)

Cells solitary or in colonies of 2-4 or 8 cells, enclosed within the old mother cell wall; ovoid to ellipsoid, thick walled and with markedly thick polar nodules. Chloroplast numerous, parietal and discoid. Cells 17.5 µ broad, 25 µ long.

## Family: Selenastraceae

## Geuns: *Selenastrum* Reinsch.

### 42) *Selenastrum bibraianum* Reinsch (Pl. 6, Fig. 1)

Philipose, (1967), p 219, fig 127-b

Free-floating colonies 4 celled, without mucilaginous envelope. Cells crescent to sickle shaped with sharply pointed ends and in colonies of 2-4-8-16 or more cells. Chloroplast single, parietal and usually with a pyrenoid. Cells 5-8µ broad, 16-38 µ long.

### 43) *S. gracile* Reinsch (Pl. 6, Fig. 2)

Rath and Adhikary, 2005, p 53, pl 7, f 17

Cells lunate to sickle shaped and quite narrow in proportion to the length, apices of cell acute, chloroplast without a pyrenoid, cells 5 µ broad, 22.5 µ long.

## Geuns: *Ankistrodesmus* Corda

### 44) *Ankistrodesmus falcatus* (Corda) Ralfs. (Pl. 6, Fig. 3)

Philipose, 1967, p 211, f 121 (a, e)

Colonies free-floating, not enclosed in mucilaginous sheath, colony of 4 cells. Cells acicular to narrowly fusiform with the ends tapering to acute apices,

Chloroplast single, parietal and usually without pyrenoids, cells 6.3 µ broad, 113.4µ long.

### 45) *A. falcatus* (Corda) Ralfs v. *radiatus* (Chod.) Lemmermann (Pl. 6, Fig. 4)

Philipose, 1967, p 211-212, fig 121 (d)

Colonies free-floating, not enclosed in mucilaginous sheath, colonies of 8 cells. Cells in radiating bundles, acicular to narrowly fusiform straight or curved with gradually tapering ends, 2.5 µ broad and 94 µ long.

### 46) *A. spiralis* (Turner) Lemmerman v. *fasciculatus* G. M.Smith (Pl. 6, Fig. 5)

Philipose, 1967, p 210-211, f 119 (d)

Colonies free-floating, not enclosed in mucilaginous sheath, cells curved or sigmoid, twisted around one another and united in colonies of 8 cells with the median portion of the cells in contact and the apices free. Cells 6 µ broad, 65 µ long, and colonies 140 µ in diameter.

## Geuns: *Kirchneriella* Schmidle

### 47) *Kirchneriella contorta* (Schmidle) Bohlin (Pl. 6, Fig. 6)

Prescott, 1951, p 258, pl 57, f 7, 8

Free floating colonies, usually of 16 twisted, arcuate, cylindrical cells with broad, convex apices, lying irregularly scattered throughout the homogeneous, gelatinous envelope; Chloroplast covering the entire wall of the cells, cells are 2.5 µ in diameter, 7.5 µ long.

### 48) *K. lunaris* (Kirch.) Moebius (Pl. 6, Fig. 7)

Prescott, 1951, p 258, pl 58, f 2

Colony composed of numerous cells arranged in groups of 4-16 within a close, gelatinous envelope; cells flat, strongly curved crescents with rather obtuse points; Chloroplast covering the convex wall; cells 5 µ broad, 10 µ long; colonies 102 µ in diameter.

## Geuns: *Actinastrum* Lagerheim

### 49) *Actinastrum hantzschii* Lagerheim (Pl. 6, Fig. 8)

Philipose, (1967), p 217, fig 125 (b)

Colonies free-floating made of 6 radially arranged cells. Cells spindle shaped and 3-6 times as long as broad. Middle of cell twice as broad as the apices. Apices attenuated or slightly rounded. Cells 3-5µ broad, 5-25µ long, colonies up to 45 µ in diameter.

## Family: Coelastraceae

### Genus: *Coelastrum* Naegeli

#### 50) *Coelastrum cambricum* Archer v. *intermedium* (Bohlin) G.S. West (Pl. 7, Fig. 1)

Prasad and Misra, 1992, p 30, pl 4, f 5

Colonies spherical, consisting of 32 cells; cells spherical in middle and sub- spherical at periphery with slightly thick, blunt and rounded projections, intercellular spaces more or less triangular; chloroplast parietal with one pyrenoid. Colony 62.5 μ in diameter, cells 15 μ in diemeter.

#### 51) *C. microporum* Naegeli (Pl. 7, Fig. 2)

Philipose, 1967, p 228, f 135 (a)

Colonies more or less spherical and of 8-16 cells with small intercellular spaces. Cells spherical to ovoid, enclosed by delicate gelatinous sheaths and interconnected by almost imperceptible gelatinous processes. Cells with sheath 6.25 μ in diameter. Colonies 35 μ in diameter.

#### 52) *C. proboscideum* Bohlin. (Pl. 7, Fig. 3)

Prasad and Misra, 1992, p 31, pl 4, f 7

Colonies pyramidal or cubical consisting of 16 cells; cells truncately conical, enclosed by a gelatinous sheath and joined along lower lateral walls, cell wall slightly thickned at poles; Chloroplast single, parietal with one pyrenoid. Colony 25 μ in diameter, cells 10 μ in diameter.

#### 53) *C. scabrum* Reinsch (Pl. 7, Fig. 4)

Philipose, 1967, p 231-132, f 140 (a)

Colony 16 celled, less spherical. Cells angular globose with three or more wart like truncate processes from the outer surface. Cells 10 μ in diameter, colony 27.5 μ in diameter.

## Family: Scenedesmaceae

### Sub Family: Crucigenioideae

### Genus: *Crucigenia* Morren

#### 54) *Crucigenia tetrapedia* (Kirchner) W.et G.S.West (Pl. 7, Fig. 5)

Philipose, 1967, p 240-241, f 151 (a)

Colony 4 celled, quadrate with a minute rectangular space at centre. Cells flattened and triangular with rounded ends. Outer side of cells always concave. Cells 10 μ in diameter.

## Sub Family: Scenedesmoideae

## Genus: *Scenedesmus* Meyen

### 55) *Scenedesmus acutiformis* Schroeder (Pl. 8, Fig. 1)

Philipose, 1967, p 260, f 169 (a)

Colony 4 celled. Cells cylindrical fusiform and arranged in a single linear series. Cell wall smooth. Median cells with a lateral longitudinal ridge extending from pole to pole on each side. Terminal cells with two or four ridges. Poles of cells acute and without teeth or spines, but sometimes with a minute papilla. Cells 7.5 μ broad, 15 μ long.

### 56) *S. arcuatus* (Lammermann) Lammermann (Pl. 8, Fig. 2)

Philipose, 1967, p 256, f 166 (j)

Colonies 8 celled, curved and with small intercellular spaces.Cells in eight celled colonies in two series, ablong avoid, sometimes slightly angular at the base due to mutual pressure. Cell wall smooth, without teeth or spines. Cells 7.5 μ broad, 17.5 μ long.

### 57) *S. arcuatus* v. *capitatus* G.M. Smith (Pl. 8, Fig. 3)

Philipose, 1967, p 257, f 166 (d-i)

Colonies curved 8-celled. Cells in colonies arranged in a double series. Cells slightly curved with one side convex and the other side straight or slightly concave. Ends of cells stumpy and with nodular thickenings. Cells 9.4μ broad, 15μ long.

### 58) *S. armatus* (Chodat) G.M. Smith v. *major* G.M. Smith (Pl. 8, Fig. 4)

Philipose, 1967, p 266, f 171 (k)

Colonies four celled. Cells oblong, ellipsoid and arranged in linear series. Terminal cells with a single long spine from each pole. All cells with a median lateral longitudinal rib which is sometimes indistinct. Four celled colony 10 μ broad, 20 μ long. Cells 7.5 μ broad, 25 μ long.

### 59) *S. armatus* (Chodat) G.M. Smith v. *bicaudatus* (Guglielmetti) Chodat (Pl. 8 Fig. 5)

Philipose, 1967, p 262, f 171 (e)

Colony 8 celled arranged in 2 series. Differs from the type in having a long spine from one of the poles of the terminal cell only, the spines of the two terminal cells alternating with each other. Cells 5 μ broad, 12.5 μ long, and spine 7.5 μ long.

### 60) *S. bijuga* (Turp.) Lagerheim (Pl. 8, Fig. 6)

Prescott, 1951, p 276, pl 63, f 2

Colony composed of 4 cells in a single flat series; cells ovate or oblong, without teeth or spine, cells 5 μ in diameter, 12.5 μ long.

### 61) *S. bijugatus* (Turpin) Kuetzing v. *flexuosus* Lemmermann (Pl. 8 and Fig. 7)

Philipose, 1967, p 254, f 164 (l)

Colony 4-8 celled with the cells arranged in irregular manner. Cells 7.5 μ broad, 8.5 μ long.

### 62) *S. bijugatus* (Turpin) Kuetzing f. *irregularis* Wille (Pl. 8, Fig. 8)

Philipose, 1967, p 253, f 164 (m)

Cells are arranged in irregular sub-alternating or sometimes in an almost double series. Inner cells often in lateral and basal contact. Colony is 8 celled. Cells 5 μ broad, 15 μ long.

### 63) *S. bijugatus* (Turpin) Kuetzing f.*parvus* (G.M. Smith) comb. nov (Pl. 8, Fig. 9)

Philipose, 1967, p 256, f 164 (h)

Colonies 4 celled, with the cells arranged in regular sub-alternating series. Cells much smaller and oblong-ovoid. Cell wall smooth. Cells 2.5 μ broad and 7.5 μ long

### 64) *S. brasiliensis* Bohlin (Pl. 8, Fig. 10)

Philipose, 1967, p 261, f 170 (a)

Colony 4 celled, cells cylindrical with attenuate apices and with longitudinal ridge extending from pole to pole on each side of the cells. Ends of cells with teeth. Cells 7.5 μ broad, 25 μ long.

### 65) *S. dimorphus* (Turpin.) Kuetzing. (Pl. 8, Fig. 11)

Philipose, 1967, p 249, f 160 (c)

Colonies 4 celled with the cells arranged in linear or subalternating series. The outer cells of the colony being more or less lunate and the apices of the cells being attenuated. Cells 7.5 μ broad, 20 μ long.

### 66) *S. dimorphus* (Turpin.) Kuetzing f. *tortus* G.M. Smith (Pl. 8, Fig. 12)

Philipose, 1967, p 251, f 160 (d)

Colony 4 celled with the cells arranged in linear series. Outer cells of colony being more lunate and the apices of the cells being attenuated. Central cells of the colony being at an angle to the axes of the terminal cells. Cell 2.5 μ broad, 20 μ long.

### 67) *S. falcatus* Chodat v. *major* Kamat (Pl. 8, Fig. 13)

Kamat, 1963, p 240, pl 9, f 42

Colony composed of 8 fusiform cells with pointed ends, arranged in alternating series, the outer cells strongly lunate, the inner cells straight; cells 8μ in diameter, 32μ long.

### 68) *S. incrassatulus* Bohlin v. *mononae* G.M. Smith (Pl. 9, Fig. 1)

Prescott, 1951, p 278, pl 63, f 13

Colony composed of 4 fusiform, subacute cells arranged in series; cells are slender; median cells slightly curved. Cells 5 μ broad, 12.5 μ long.

### 69) *S. indicus* Philipose (Pl. 9, Fig. 2)

Philipose, 1967, p 258, f 167

Colony 4 celled. Cells curved with capped ends and swollen middle and arranged in a sub-alternating series with the outer ends of internal cells in a contact with the middle of the nearest terminal cell, the inner ends remaining free. Internal cells also in contact with each other at the median region. Cells 5 μ broad, 12.5 μ long.

### 70) *S. longus* Meyen v. *naegelii* (Brebisson) G.M. Smith (Pl. 9, Fig. 3)

Philipose, 1967, p 274, f 180 (c)

Colony 2 celled, cells more or less oblong to cylindrical to subpyriform with rounded ends, and arranged in linear series. Terminal cells with a long recurved spine from one pole and a long or short, straight or slightly curved spine from the other pole. Internal cells with a short erect spine usually from one pole only, but sometimes with rudiments of spine from the other pole.

### 71) *S. obliquus* (Turpin) Kuetzing (Pl. 9, Fig. 4)

Philipose, 1967, p 248, f 159 (a)

Colony composed of 4 cells arranged in linear series. Cells fusiform with acute or slightly rounded cells and usually with straight sides. Outer side of terminal cell concave or slightly convex. Cell wall smooth and without terminal teeth or spines. Cells 5 μ broad, 12.5 μ long.

## 72) *S. opoliensis* P. Richter (Pl. 9, Fig. 5)

Philipose, 1967, p 275, f 181 (b)

Colonies 2-celled with subfusiform, cells arranged in a linear series. Adjacent cells in contact only along about a third of their length. Terminal cells often narrower and subrectangular. Poles of terminal cells with a long, more or less recurved spine. Cells 5.2 μ broad, 19.3 μ long, spines 14.6 μ long

## 73) *S. opoliensis* v. *mononensis* Chodat (Pl. 9, Fig. 6)

Philipose, 1967, p 276, fig 181 (c)

Colony eight celled. Internal cells broadly fusiform with pointed ends, without spine from their poles. Terminal cells with attenuate, semitruncate to rostrate ends and with a long, straight or recurved spine from each pole. Cells 5 μ broad, 22 μ long. Spines of terminal cells 18.9μ long.

## 74) *S. protuberans* Fritschet. Rich f. *minor* Ley (Pl. 9, Fig. 7)

Philipose, 1967, p 277, f 182

Colonies usually four-celled. Cells in a linear series and laterally in close contact with adjoining cells except at the ends. Terminal cells longer than the inner cells with their apices drawn out and protruding, and with a long spine usually arising from the outer side of each end. Inner cells with pointed or slightly truncate ends. Cells 5μ broad, 18.9μ long.

## 75) *S. quadricauda* v. *bicaudatus* Hansg. (Pl. 9, Fig. 8)

Philipose, 1967, p 284, f 187 (k)

Colonies 4-celled. Terminal cells with a long spine from one pole only, the spine of one terminal cell being at an angle opposite to that of the other terminal cell. Internal cells without spines from their poles. Cells 4-5 μ broad, 8-12 μ long.

## 76) *S. platydiscus* (G.M. Smith) Chodat (Pl. 9, Fig. 9)

Philipose, 1967, p 256, f 165

Colony flat, 8 celled, with oblong elliptic cells arranged in a double series. Interstices between the cells minute or absent. Cells 5μ broad, 10μ long.

## 77) *S. quadricauda* (Turpin) Brebisson v. *longispina* (Chodat) G.M. Smith (Pl. 9, Fig. 10)

Philipose, 1967, p 285, f 187 (c)

Colonies usually 4 celled, rarely 8 celled. Cell ovoid to cylindrical with the cells narrower than in the type and the spines proportionately longer, compared to the length of the cells. Cell 3.7 μ broad, 12.6 μ long.

**78) *S. quadricauda* (Turpin) Brebisson v. *maximus* West and West (Pl. 9, Fig. 11)**

Prescott, 1951, p 280, pl 64, f 4

Colony composed of 4 oblong cylindrical cells in one series; outer cell with a long curved spine at each pole; inner cells without spines; cells 10 µ in diameter, 27.5 µ long; spines 17.5 µ long.

**79) *S. quadricauda* (Turpin) Brebisson v. *quadrispina* (Chodat) G.M. Smith (Pl. 9, Fig. 12)**

Philipose, 1967, p 285, f 187 (d)

Colonies two celled, cells broadly ovoid and about twice as long as broad. Cell wall smooth. Poles of terminal cells with a single short recurved spines. Cells 7.8 µ broad, 15 µ long and spine 5 µ long.

**80) *S. quadricauda* v. *westii* G.M. Smith (Pl. 9, Fig. 13)**

Philipose, 1967, p 286, f 187 (i)

Colonies is composed of 8 cells. Cells 10 µ broad, 12.5 µ long, and Spines 12.5 µ long.

## Order: Ulothrichales

## Family: Ulothrichaceae

### Genus: *Uronema* Lagerheim

**81) *Uronema confervicolum* Lagerheim (Pl. 10, Fig. 1)**

Prasad and Misra, 1992, p 46, pl 6, f 3

Filaments long, slightly curved, consisting of many cells, constricted at septa; cells cylindrical, slightly longer than broad, apical cells acuminating with pointed apex, basal cell narrow and long; each cell with one laminate chloroplast, accupying a part of the cell; pyrenoid not prominent.

### Genus: *Ulothrix* Kuetzing

**82) *Ulothrix subtilissima* Rabenhorst (Pl. 10, Fig. 2)**

Prescott, 1951, p 96, pl 6, f 3

Filaments long and slender, free floating or attached, cells very slightly inflated and constricted at the cross walls. Chloroplast extending the entire length of the cell, with one pyrenoid. Cells 5 µ in diameter, 12.5 µ long.

**83) *U. tenerrima* Kuetzing (Pl. 10, Fig. 3)**

Prescott, 1951, p 96, pl 6, f 12

Filaments free floating or attached; long or in short sections; composed of cylindrical, relatively short cells with constriction at the length of cell, with one pyrenoid. Cells 7.5 μ in diameter, 12.5 μ long.

### 84) *U. tenuissima* Kutzing (Pl. 10, Fig. 4)

Ramnathan, 1964, p 34, pl 9, f (g)

Cells mostly 15-22 μ thick, in young stages as long as broad, but most often ¼-½ the breadth; cell wall thin, not striated; chloroplast a broad band, with two or more pyrenoids; zoospores formed in somewhat swollen cells.

### 85) *U. variablis* Kuetzing (Pl. 10, Fig. 5)

Prescott, 1951, p 97, pl 6, f 13

Filaments long, slender and entangled, forming cottony masses. Cells cylindrical without constrictions at the cross walls, chloroplast folded, parietal plate with one pyrenoid. Cells 5.5 μ in diameter and 15 μ long.

### 86) *U. zonata* (Weber *et* Mohr) Kuetzing (Pl. 10, Fig. 6)

Ramnathan, 1964, p 30, pl 5, f (M)

Filaments attached in earlier stages, later free, with cylindrical or slightly swollen cells, of very varying width and length; length generally smaller than width except in some cases when it can be 2-3 times width; cells measures generally 11-45 μ broad and 1/3- $1^{1/2}$ times as long, cell wall thin in younger stages, but thickened in older filaments; chloroplast band shaped, broad or narrow girdle, covering only the medium region of the cell and containing more than one pyrenoid.

## Genus: *Geminella* Turpin emend. Lagerheim

### 87) *Geminella minor* (Naegeli) Heering (Pl. 10, Fig. 7)

Ramnathan, 1964, p 65, pl 18, f (k)

Cells cylindrical, 10 μ broad, 25 μ long, always united together and surrounded by a mucilaginous sheath 7.5 μ thick; chloroplast parietal, covering part of the cell, with one or more pyrenoid.

## Family: Schizomeridaceae

## Genus: Schizomeris Kuetzing

### 88) *Schizomeris leibleinii* Kuetzing (Pl. 10, Fig. 8)

Ashtekar, 1980, p 31, pl 10 fig 40

Filaments uniseriate below with cylindrical cells, multiseriate above with brick like cells, 20-24 μ in diameter below, and as much as 100 μ wide in the multiseriate, upper portion of the frond; cells 10-20 μ in diameter.

## Order: Cladophorales

## Family: Cladophoraceae

### Genus: *Rhizoclonium* Kuetzing

**89) *Rhizoclonium hieroglyphicum* (Ag.) Kuetz. (Pl. 10, Fig. 9)**

Prasad and Misra, 1992, p 56, pl 7, f 7

Filaments long, wiry, unbranched at septa; cells long, cylindrical, with thick stratified walls, multinucleate, chloroplast parietal, reticulate with several pyrenoids; rhizoids primary, long and colourless, formed from the direct prolongation of the lower and of filaments. Cells 27.5 μ broad, 132.5 μ long.

### Genus: *Cladophora* Kuetzing

**90) *Cladophora fracta* (Dillw.) Kuetzing (Pl. 11, Fig. 1)**

Prescott, 1951, p 137, pl 20, f 3

Floating, forming coarse, light green masses of irregularly branched filaments, the branches often curving. Cells irregularly swollen or clavate; 65 μ in diameter in main axis, 1-3 times their diameter in length; 25 μ diameter in ultimate branches, 3-6 times their diameter in length.

**91) *C. fracta* v. *lacustris* (Kuetz.) Brand *ex* Heering (Pl. 11, Fig. 2)**

Prescott, 1951, p 138, pl 20, f 7

A coarse, rigid form with very little branching, the branches arising irregularly, often at right angles to the main axis. Cells long and cylindrical, 42.5 μ in diameter.

**92) *C. insignis* (C.A.Ag.) Kuetzing (Pl. 11, Fig. 3)**

Prescott, 1951, p 139, pl 21, f 5

Filaments floating, thallus composed of straight, coarse and sparsely branched filaments, the branches often arising at right angles to the main axis and very slightly smaller. Main filament 77.5 μ in diameter, the cells cylindrical or swollen; branches 52.5 μ in diameter.

### Genus: *Pithophora* Wittrock

**93) *Pithophora oedogonia* (Mont.) Wittrock (Pl. 11, Fig. 4)**

Prasad and Misra, 1992, p 55, pl 7, f 3

Filaments slender, coarse, branching mostly solitary, cells cylindrical, longer than broad; cell wall thin; akinetes intercalary and terminal, intercalary akinetes swollen, oblong to globose, thick walled, 175 μ long, 95 μ broad.

### Genus: *Basicladia* Hoffman and Tilden

#### 94) *Basicladia chelonum* (Collins) Hoffman and Tilden (Pl. 11, Fig. 5)

Prescott, 1951, p 143, pl 23, f 11

Filament a coarse, attached, errect filament; branching only at the base, just above the attaching organs, which anchor the plant to the backs of turtles. Main filament 17.5 µ in diameter in the distal region. Cells cylindrical, especially the basal cell which may be as much as 50 times longer in length. Walls thick and lamellate.

## Order: Oedogoniales

### Family: Oedogoniaceae

### Genus: *Oedogonium* Link

#### 95) *Oedogonium capilliforme* Kuetzing v. *australe* Wittrock *in* Wittrock and Nordstedt (Pl. 12, Fig. 1)

Prescott, 1951, p 168, pl 29, f 5

Macrandrous; dioecious, vegetative cells cylindrical, 32.5 µ in diameter, 67.5 µ long. Oogonia solitary; obvoid, globose; opening by a superior pore 52.5 µ broad, 55 µ long. Oospores subglobose; not filling the oogonia; wall thick, smooth; 42.5 µ in diameter, 45 µ long

#### 96) *O. cardiacum* (Hass.) Wittrock (Pl. 12, Fig. 2)

Prescott, 1951, p 168, pl 29, f 7

Macrandrous; diocious, vegetative cells cyllindrical, 20 µ in diameter, 52.5 µ long. Oogonia solitary, globose; opening by supramedian pore 32.5 µ in diameter, 27.5 µ long. Oospores globose, not filling the oogonia; cell wall thick, smooth.

#### 97) *O. cymatosporum* Wittr. Nordst. (Pl. 12, Fig. 3)

Gonzalves, 1981, p 195, f 9.65 (A)

Macrandrous, homothallic; vegetative cells cylindrical, 7.5 µ in diameter, 32.5 µ long; oogonium single, subdepressed-globose, 22.5 µ in diameter, 25 µ long; spore wall three layered, outer and inner layer smooth, median layer scrobiculate.

#### 98) *O. gallicum* Hirn. (Pl. 12, Fig. 4)

Prasad and Misra, 1992, p 70, pl 11, f 13-15

Nannandrous, idioandrosporous; vegetative cells cylindrical, 19 µ broad, 59.6 µ long, oogonium solitary, globose, pore superior, 44 µ broad, 50 µ long; mesospore thick smooth, brownish yellow.

### 99) *O. glabrum* f. *glabrum* Hall. (Pl. 12, Fig. 5)

Dhande and Jawale, 2006, p 321, f 1, 2

Marandrous, heterothallic; vegetative cells cylindrical, those of female filaments 21 µ in diameter, 63 µ long, ogonium single, obovoid, often terminal, 52 µ in diameter, 67 µ long, poriferous, pore superior

### 100) *O. globosum* Nordst. (Pl. 12, Fig. 6)

Prasad and Misra, 1992, p 69, pl 11, f 12

Macrandrous, monoecious; vegetative cells cylindrical 10 µm broad, 25 µm long; oogonium solitary, globose to faintly sub globose with superior pore, 30 µm in diameter; mesospore smooth; antheredia usually 2, rarely upto 4, subhypogynous with two superimposed antherozoids.

### 101) *O. kufferathii* Gauth. (Pl. 12, Fig. 7)

Jawale and Dhande, 2005, p 19, f 9

Marandrous, heterothallic; vegetative cells cylindrical, those of female filaments 21 µ in diameter, 82.6 µ long, terminal cell obtuse; oogonium single, ellipsoid, more elongate, 54 µ in diameter, 67.3 µ long, operculate, division superior.

### 102) *O. patulum* Tiffany (Pl. 12, Fig. 8)

Prescott, 1951, p 180, pl 33, f 8

Vegetative cells capitellate, 15 µ in diameter, 47.5 µ long. Oogonia solitary; globose to sub-globose; opening by a superior pore, 40 µ in diameter; 40 µ long. Oospore globose; cell wall smooth.

### 103) *O. pseudotumidulum* Gauth.- Liev. (Pl. 12, Fig. 9)

Jawale and Dhande, 2005, p 19, f 10

Macrandrous, heterotallic; vegetative cells cylindrical, 24.5 µ in diameter, 75µ long, terminal cell obtuse; oogonium single, subellipsoid, 55 µ in diameter, 62µ long, operculate, division superior, oospore ellipsoid, 50.5 µ in diameter, 57 µ long, spore wall thick, longitudinally costate, costae about 13, undulate connected by fine regularly placed transverse lines.

## Order: Zygnematales

## Family: Zygnemataceae

## Genus: *Mougeotia* C.A.Agardh

### 104) *Mougeotia drouetii* Transeau (Pl. 13, Fig. 1)

Randhawa, 1959, P128, f 18

Thallus green, free-floating. Vegetative cells 12-14 x 150-160 μ; chloroplast with 4-8 pyrenoids in single row. Conjugation scalariform; zygospore globose in the enlarged tube, extending into gametangia.

### 105) *M. calcarea* (Cleve) Wittrock (Pl. 13, Fig. 2)

Randhawa, 1959, p 123, f 10

Vegetative cells 100 μ long; chloroplast with 4-8 pyrenoids in a single row; cells elongating. Conjugation scalariform; zygospore formed whooly in the conjugation tube or extending in to one or both gametangia, globose, or angular globose; spore wall smooth, colorless; aplanospores globose, lateral to sporogenous cell, or rarely trapezoid-ovoid, dividing the sporogenous cells.

### 106) *M. scalaris* Hassall. (Pl. 13, Fig. 3)

Prescott, 1951, p 304, pl 71, f 7

Vegetative cells 30 μ broad, 77.5 μ long, and chloroplast broad plate with 4 pyrenoids. Zygospores formed in the tube by scalariform conjugation.

### 107) *M. sphaerocarpa* Wolle (Pl. 13, Fig. 4)

Prasad and Misra, 1992, p 75, pl 12, f 7

Vegetative cells 20 μ broad, 125 μ long; chloroplasts plate like with 9 pyrenoids, conjugation scalariform by tubes formed by both gametangia; zygospores formed in greatly enlarged and broad conjugating tubes and extending somewhat in both gametangia, ovoid to subglobose in shape; median spore wall smooth, brown.

## Genus: *Zygnema* C.A. Agardh

### 108) *Zygnema khannae* Skuja (Pl. 13, Fig. 5)

Prasad and Misra, 1992, p 76, pl 12, f 9

Vegetative cells 22.5 μ broad, 65 μ long; conjugation not seen; reproduction by aplanospores; aplanospores ellipsoid to ovoid; outer spore wall punctate, median spore wall irregular, yellowish brown in colour.

### 109) *Z. pectinatum* (Vauch.) C.A. Agardh (Pl. 13, Fig. 6)

Prescott, 1951, p 325, pl 69, f 9

Filaments forming light green, cottony masses; vegetative cells 32.5 µ in diameter, 67.5 µ long. Zygospores forms in tubes.

### 110) *Z. synadelphum* Skuja (Pl. 13, Fig. 7)

Prescott, 1951, p 326, pl 74, f 1

Vegetative cells 20 µ in diameter; fertile cells not inflated. Zygospores formed in the tube; globose, parallel with the conjugation tube; madian layer blue and scrobiculate; 35 µ in diameter.

## Genus: *Zygnemopsis* (Skuja) Transeau

### 111) *Zygnemopsis jogensis* Iyengar (Pl. 13, Fig. 8)

Randhawa, 1959, p 208, f 140 (a)

Conjugation scalariform, vegetative cells 17.5 µ broad, 27.5 µ long; zygospores lying in the widened conjugation canal and extending far into each gametangium; zygospores irregularly cushion shaped, 40 µ broad, 45 µ long; chloroplast two, with one pyrenoid in each.

### 112) *Zygnemopsis stephensiae* Transeau. (Pl. 13, Fig. 9)

Randhawa, 1959, p 195, f 125

Vegetative cells 12-14 x 60-76 µ with 2 more or less stellate chloroplasts. Conjugation scalariform; zygospore formed in the greatly enlarged conjugating tubes and extending into both gametangia, spores quadrately ovoid, 29-32 x 42-45 µ; with angles rounded, and sides concave; irregularly scrobiculate.

## Genus: *Debarya* (Wittr.) Transeau

### 113) *Debarya costata* Randhawa (Pl. 13, Fig. 10)

Randhawa, 1959, p 109, f 5 (a)

Vegetative cells 10 µ broad and each cell has a single plate shaped chloroplast bearing six pyrenoids. Conjugation scalariform. The conjugation processes are given out in the shape of long cylindrical phallic bodies, which meet and produce long conjugation canals.

## Genus: *Spirogyra* Link

### 114) *Spirogyra aplanospora* Randhawa (Pl. 14, Fig. 1)

Randhawa, 1959, p 370, f 402

Filament green, free-floating, vegetative cells 38 µ broad, and 2-4 times as long. Septa plane. Each cell contains a single spiral chloroplast with 3-6 spirals.

Reproduction in this species takes place exclusively by means of aplanospores. In most filaments, the fertile cells become swollen on both sides, appearing almost globose in shape, while in some filaments these become swollen on one side only. Abortive conjugation tubes are given out by some of the cells, presenting the appearance of a broken conjugation canal. The aplanospores are globose to ellipsoid in shape and are 36 μ broad and 38 μ long. The spore wall is composed of two layers.

### 115) *Spirogyra borgeana* Transeau (Pl. 14, Fig. 2)

Prescott, 1951, p 311, pl 77, f 8

Filaments of cylindrical cells, 27.5 μ in diameter, and 175 μ long, with plane end walls. Chloroplast solitary, making $1^{1}/_{2}$ to 5 turns. Conjugatiom scalariform; sporangia inflated only on the outer side.

### 116) *S. collinsii* (Lewis) Printz (Pl. 14, Fig. 3)

Prescott, 1951, p.312, pl.77, f 6

Filaments of rather slender cells, 20 μ in diameter and 105 μ long, with plane end walls. Chloroplast solitary. Conjugation scalariform and lateral, the tube formed by male gametangium only; sporangia inflated slightly on both sides to contain the spores. Zygospores ellipsoid; median wall coarsely punctate.

### 117) *S. crassa* Kuetzing v. *minispora* Prasad and Misra (Pl. 14, Fig. 4)

Prasad and Misra, 1992, p 79, pl 13, f 1

Vegetative cells 100 μ broad, 122.5 μ long, end wall plane; each cell with 6 chloroplasts, making half turn; conjugation scalariform, tubes formed by both gametangia; zygospores oval to rounded; median wall smooth, yellowish.

### 118) *S. condensata* (Vaucher) Kuetzing (Pl. 14, Fig. 5)

Prasad and Misra, 1992, p 78, pl 13, f 12

Vegetative cells 47.5 μ broad, 80 μ long, end walls plane; each cell with single chloroplast, making 2.5-4 turns; conjugation lateral; fertile cells almost cylindric; zygospores forming cells usually in pair; median spore wall smooth, yellow.

### 119) *S. fallax* (Hansg.) Wille (Pl. 14, Fig. 6)

Prescott, 1951, p 314, pl 77, f 10

Filaments of rather slender cylindrical cells, 35 μ in diameter and up to 8 times the diameter in length, with replicate end walls, conjugation scalariform, the tubes short but formed from both gametangia; fertile cells inflated. Zygospores ellipsoid; wall layer smooth; 45 μ in diameter.

## 120) *S. gracilis* (Hassall) Kuetzing (Pl. 14, Fig. 7)

Prasad and Misra, 1992, p 83, pl 14, f 3

Vegetative cells 17.5 µ broad, 65 µ long, end walls plane; each cell with one chloroplast making 2 to3.5 turns; conjugation scalariform by tubes from both gametangia; fertile cells inflated on conjugation side; zoospore ellipsoid with rounded ends. Median spore wall yellow, smooth with distinct suture.

## 121) *S. gratiana* Transeau (Pl. 14, Fig. 8)

Prescott, 1951, p 315, pl 74, f 9

Filaments of elongate cylindric cells, 30 µ in diameter, 155 µ long, with replicate end walls; Chloroplasts two to three. Conjugation lateral or scalariform by tubes from both gamentagia; fertile cells cylindric, or merely enlarged. Zoospores ellipsoid; median spore wall smooth and yellow.

## 122) *S. hyalina* Cleve (Pl. 14, Fig. 9)

Randhawa, 1959, p 318, f 294 (a)

Vegetative cells 60 x 82.5 µ, with plane end walls, chloroplast three or four, conjugation scalariform or lateral, tubes formed by both gametangia; fertile cells cylindric; zygospores ellipsoid, more or less pointed 47.5 µ broad, 60 µ long; median spore wall brown and smooth.

## 123) *S. irregularis* Naegeli (Pl. 14, Fig. 10)

Prasad and Misra, 1992, p 84, pl 14, f 9

Vegetative cells 40 µ broad, 175 µ long, end wall plane; each cell with 4 chloroplasts making 2 to 3 turns; conjugation scalariform by tubes formed by both gametangial cells.

## 124) *S. juergensii* Kuetz. (Pl. 15, Fig. 1)

Randhawa, 1959, p 294, f 250

Thallus green, free-floating, vegetative cells 25-27 µ broad, 108µ long, with plane end walls; chloroplast 1 to 2 making 2-4 turns in the cell. Conjugation scalariform.

## 125) *S. neglecta* (Hassall) Kuetzing (Pl. 15, Fig. 2)

Prasad and Misra, 1992, p 85, pl 14, f 10

Vegetative cells 50 µ broad, 160 µ long, end walls plane; each cell with 3 chloroplasts making 2.5 - 3 turns; conjugation scalariform by tubes formed by both gametangial cells; receptive gametangia slightly swollen.

## 126) *S. orientalis* W. and G.S. West (Pl. 15, Fig. 3)

Randhawa, 1959, p 333, f 324

Thallus green, free-floating, vegetative cells 36-38 µ broad; 84-98 µ long; end walls plane; chloroplast 3, making 1 to 1.5 turns. Conjugation scalariform; tubes formed by both gametangia; fertile cells inflated, zygospore ellipsoid, 38-42 µ broad, 60 µ long; median spore wall minutely scrobiculate, brown.

### 127) S. *parvispora* Wood (Pl. 15, Fig. 4)

Prasad and Misra, 1992, p 86, pl 14, f 12

Vegetative cells 75 µ broad, 200 µ long, end wall plane; each cell with 4 chloroplasts making 2 turns; conjugation scalariform by both gametangial cells; fertile cells not swollen; zygospores ellipsoid; mesospore smooth, brown.

### 128) S. *parvula* (Trans.) Czurda (Pl. 15, Fig. 5)

Randhawa, 1959, p 298-299, f 259

Filament green, free-floating, vegetative cells 24-28µ broad; 2-5 times as long. Each cell with one chloroplast of 2-4 turns. Septa plane. In scalariform conjugation the female cells become swollen. The zygospore ellipsoid to oval in shape 34-36 µ broad, 60-65 µ long.

### 129) S. *plena* (W. and G.S. West) Czurda (Pl. 15, Fig. 6)

Randhawa, 1959, p 325-326, f 310

Thallus green, free-floating, vegetative cells 45 x 57µ, with plane end walls; 2 to 3 chloroplasts. Conjugation scalariform and lateral; tubes large and formed by both gemetangia; fertile cells cylindric; zygospore ovoid, 42 x 68 µ; median spore wall yellow.

### 130) S. *pratensis* Transeau (Pl. 15, Fig. 7)

Prescott, 1951, p 319, pl 75, f 5

Filaments of rather slender cells, 17.5 µ in diameter, 95 µ long, with plane end wall, Chloroplast solitary, making 1-8 turns, conjugation by tubes from both gametangia; fertile cells inflated. Zygospores ellipsoid to subcylendric-ovate; median spore wall smooth, yellow at maturity.

### 131) S. *propria* Transeau (Pl. 15, Fig. 8)

Randhawa, 1959, p 418

Vegetative cells 62.5 µ broad, 80 µ long with plane end walls; chloroplasts three making 0.5 to 1 turn. Conjugation lateral; tubes formed by both gametangia; fertile cells cylindric; zygospore ellipsoid, 42.5 µ broad, 60 µ long; median spore wall irregularly pitted.

### 132) S. *pseudofloridana* Prescott (Pl. 15, Fig. 9)

Prescott, 1951, p 319, pl 75, f 7-9

Thallus green, free floating, filaments of stought cells, 51-60 μ in diameter and 2-5 times the diameter in length, with plane end walls; chloroplast 4, narrow making ½ and 1 ½ loose turnes. Gametangia becoming shortened and apparently cut off from vegetative cells, fertile cells slightly inflated zygospore ellipsoid, 48-62 μ in diameter, 72-100 μ long.

## 133) S. *pulchrifigurata* Jao (Pl. 15, Fig. 10)

Randhawa, 1959, p 334, f 328 (a)

Vegetative cells 50 x 65 μ, with plane end walls; 3 chloroplast making 4 turns, conjugation scalariform; tubes formed by both gametangia; fertile cells inflated up to 83 μ. Zygospores ellipsoid with more or less rounded ends, 50 x 70 μ; median spore wall yellow, brown and irregularly reticulate.

## 134) S. *rhizobrachialis* Jao (Pl. 16, Fig. 1)

Randhawa, 1959, p 336, f 331, Prescott, 1951, p 320, pl 76, f 1, 2

Vegetative cells, 57.6 x 129.6 μ; end walls plane; chloroplast 3-5, in some portions of the filament each sterile cell produces a lateral rhizoid-shaped attachment with a more lobed hapteron.

Conjugation scalariform; tubes formed by both gametangia; fertile cells cylindric, sometimes shortened and enlarged; zygospores ellipsoid with more rounded ends, 50 x 72 μ; median spore wall yellow brown.

## 135) S. *rivularis* (Hassal) Rabenhorst (Pl. 16, Fig. 2)

Prasad and Misra, 1992, p 88, pl 14, f 4,6

Vegetative cells 42.5 μ broad, 125 μ long, end walls plane; each cell with three chloroplasts making 2-2.5 turns; conjugation scalariform by tubes formed by both gametangial cells; fertile cells cylindrical, zygospores ellipsoid 45 μ broad, 75 μ long; median spore wall smooth and yellowish brown.

## 136) S. *setiformis* (Roth) Kg. (Pl. 16, Fig. 3)

Sen and Naskar, 2003, p 133, pl 12, f 74

Thallus free-floating nearly brown, cells of the filaments regularly cylindrical usually of a size throughout the filament, massive filaments, cells 102.5 μ broad, 105 μ long, chloroplast 4-6, running parallel in angular strips.

## 137) S. *spreeiana* Rabenhorst (Pl. 16, Fig. 4)

Prescott, 1951, p 321, pl 77, f 9

Vegetative cells 20 μ in diameter and up to 25 times the diameter in length; chloroplast solitary, making 1½ to 4 turns Conjugation tubes formed from the male gametangia only; fertile cells inflated on both sides. Zygospores ellipsoid, median spore wall smooth and yellowish; 30 μ in diameter, 55 μ long.

### 138) S. *stictica* (Engl.Bot) Wile (Pl. 16, Fig. 5)

Prescott, 1951, p 321, pl 76, f 5

Filaments of long cells, 45 µ in diameter, 292.5 µ long, with plane end walls; Chloroplast 2-6, straight or making 1/2turn, conjugation by geniculate bending of filaments; connecting tubes are not formed; fertile cells becoming slightly swollen. Zygospore ellipsoid; median spore wall smooth.

### 139) S. *subsalsa* Kuetzing (Pl. 16, Fig. 6)

Prescott, 1959, p 321, pl 73, f 1-3

Vegetative cells 40-50 µ broad and 139.2 µ in length, with plane end walls; solitary, chloroplast making 2 turns. Conjugation scalariform; conjugation by tubes from both the gametangia. The specimen is slightly broader than the type described by Prescott (1959).

### 140) S. *triplicata* (Collins) Transeau (Pl. 16, Fig. 7)

Randhawa, 1959, p 324, f 307

Vegetative cells 36 x 84 µ; end walls plane; chloroplast 3, making 1.5 to 3 turns. Conjugation scalariform; tubes formed by both the gametangia; fertile cells cylindrical zygospores ovoid.

### 141) S. *undulisepta* Randhawa (Pl. 16, Fig. 8)

Randhawa, 1959, p 353-354, f 367

Vegetative cells are 19.2 µ broad, and 8-12 times as long, each containing a single irregularly spiral chloroplast bearing 3-8 pyrenoids. The septa are peculiarly replicate and look like a parallelogram with the middle lamella as a diagonal, and the longer sides undulating in a wave-like manner.

### 142) S. *variformis*Transeau. (Pl. 16, Fig. 9)

Randhawa, 1959, p 381-382, f 427

Thallus green, free-floating. Vegetative cells 69.6 x 130 µ with plane end walls; 1 chloroplast making 2 to 5 turns in the cell. Conjugation scalariform, zygospore 40-45 µ.

### 143) S. *weberi* Kuetzing (Pl. 16, Fig. 10)

Prescott, 1951, p 322, pl 76, f 8

Filaments of long cells, 27.5 µ in diameter, and 100 µ long, with replicate end walls; chloroplast solitary, broad, making 3 to 6½ turns. Conjugation by tubes from both gametangia; fertile cells cylindrical. Zygospores cylindric-ovate; median spore wall smooth and brown.

## Family: Mesotaeniaceae

### Genus: *Cylindrocystis* Meneghini

**144) *Cylindrocystis subpyramidata* W. *et* G.S. West (Pl. 17, Fig. 1)**

Prasad and Misra, 1992, p 89, pl 15, f 11

Cells cylindrical, about 1-5 times longer than broad, slightly constricted in middle, cell apices subpyramidal with rounded ends; chloroplast substellate with one large pyrenoid in each semicell. Long cell 27.5 μ, lat. Cell 15 μ.

## Family: Desmidiaceae

### Genus: *Closterium* Nitzsch

**145) *Closterium acerosum* (Schrank) Ehr. v. *elongatum* Breb. (Pl. 17, Fig. 2)**

Prasad and Misra, 1992, p 98, pl 16, f 20

Cells are much longer, inner margin more or less straight, outer margin moderately curved, cell apices tapering with rounded to subacute ends; cell wall with fine striations; chloroplast ridged, containing pyrenoids in median series. Long cell 750 μ, lat. cell 40.5 μ, lat. apex 1.5 μ.

**146) *C. acutum* (Lyngb.) Breb. (Pl. 17, Fig. 3)**

Prasad and Misra, 1992, p 99, pl 16, f 4

Cells small, somewhat curved, 18-21 times longer than broad, outer margin 55-70 degrees of arc, inner margin not tumid; gradually attenuated to acute apices; cell wall smooth; chloroplast with 6-8 pyrenoids in a central series. Long cell 72-83μ, lat. cell 3.75-4.5μ

**147) *C. calosporum* Witr. v. *maius* W.*et* G.S.West. (Pl. 17, Fig. 4)**

Prasad and Misra, 1992, p 102, pl 16, f 24

Cells larger, outer margin strongly curved, inner margin parallel to outer near apices and somewhat straight in the middle, cell gradually attenuated to subacute apices; cell wall smooth; chloroplast with single series of pyrenoids. Long. cell 122.5 μ, lat. cell 15.5 μ, lat.apices 2.5 μ.

**148) *C. dianae* Ehrenberg (Pl. 17, Fig. 5)**

Prasad and Misra, 1992, p 105, pl 16, f 7

Cells medium size, 9-10 times longer than broad, outer margin strongly curved with 112-125 degree of arch, inner slightly tumid, cell gradually attenuated towards obtusely rounded apex; cell wall smooth; chloroplast ridged with 6-8 pyrenoids in row.Long.cell 285 μ, lat.cell 30 μ, lat.apex 5 μ.

### 149) *C. incurvum* Breb. (Pl. 17, Fig. 6)

Prasad and Misra, 1992, p107, pl 16, f 8

Cells small, 7-8 times longer than broad, strongly curved, outer margin with arc, inner margin concave but not tumid in the middle, cell attenuated towards acute apices; cell wall smooth; chloroplast ridged with four pyrenoids arranged in row. Long. cell 95 μ, lat. cell 12.5 μ, lat. apex 1.5 μ.

### 150) *C. kuetzingii* Breb. (Pl. 17, Fig. 7)

Prasad and Misra, 1992, p 107, pl 16, f 22

Cells of medium size, longer than broad, almost straight, medium part fusiform lanceolate with convex margins, cells tapering towards each extremity and ending in long setaceous processes with parallel sides and rounded apices; cell wall longitudinally striated, striae delicate, visible across the cell; chloroplast with 4 pyrenoids, arranged in row. Long. cell 430 μ, lat. cell 22.5 μ, lat. apex 2.5 μ.

### 151) *C. leibleinii* Kuetz. (Pl. 17, Fig. 8)

Misra, Mehrotra, Jai Prakash and Srivastava, 2001, p 4, pl 2, f 6

Cells of medium size, 7-8 times longer than broad, strongly curved, outer margin with 154-180 degrees of arc, inner margin concave and slightly tumid in the middle, cell gradually attenuated to the acutely rounded apices, cell wall smooth, chloroplast with 4-5 ridges and a median row of 8-9 pyrenoids, long cell 229-235μ, lat. cell 31-37μ, lat. apex 7μ.

### 152) *C. libellula* Focke v. *intermedium* (Roy et Biss.) G.S. West (Pl. 17, Fig. 9)

Prasad and Misra, 1992, p 110, pl 17, f 16

Cells of medium size, 5-6 times longer than broad, fusiform, more or less straight with both the margins moderately convex, cell gradually attenuated from middle towards broadly rounded and subtruncate poles; cell wall smooth; chloroplast with 5-7 ridges and 8-9 pyrenoids in an axile row. Long cell 139-142.5μ, lat. cell 24.5-27μ, lat. apex 10-12μ.

### 153) *C. moniliferum* (Bory) Ehrenberg (Pl. 18, Fig. 1)

Prasad and Misra, 1992, p 113, pl 17, f 5

Cells of medium size, stout, 6-7 times longer than broad, moderately curved, outer margin 120-130 degree of arc, inner margin inflated in the middle, cell uniformly narrowed to obtusely rounded apices; Cell wall smooth; chloroplast with 4-5 ridges and 8-10 pyrenoids arranged in median series. Long. cell 215 μ, lat. cell 35 μ, lat. apex 7.5 μ.

### 154) *C. parvulum* Naeg. (Pl. 18, Fig. 2)

Prasad and Misra, 1992, p114, pl 16, f 17

Cells small, 10-12 times longer than broad, strongly curved, outer margin with125-130 degree of arch, inner margin equally curved, not tumid in the middle, cell gradually attenuated to acutely rounded apices; cell wall smooth; chloroplast with ridges and an axile row of 6 pyrenoids. Long. cell 122.5 μ, lat. cell 13 μ, lat. apex 3 μ.

### 155) *C. parvulum* Naeg. v. *angustum* W.et G.S. West (Pl. 18, Fig. 3)

Prasad and Misra, 1992, p115, pl 16, f 16

Cells relatively smaller, narrower and less curved, outer margin with 125-130 degree arc, inner margin equally curved, not tumid in middle, cells gradually attenuated to acutely rounded apices; cell wall smooth, chloroplast with ridges and axile row of pyrenoids. Long. cell 92.5 μ, lat. cell 7.5 μ.

### 156) *C. peracerosum* Gay (Pl. 18, Fig. 4)

Misra, Mehrotra, Jai Prakash and Srivastava, 2001, p 4, pl 2, f 5

Cells of medium size, 7-8 times longer than broad, moderately curved, outer margin 30 degrees of arc, inner slightly concave towards poles and more or less straight in mid region, cell gradually attenuated towards acute apices, cell wall smooth, chloroplast ridged with an axial series of pyrenoids. Long cell 192-199μ, lat. cell 14-19μ, lat. apex 3-4μ.

### 157) *C. rectimarginatum* Scott and Prescott. (New variety) (Pl. 18, Fig. 5)

Scott and Prescott, 1961, p 13, pl 1, f 28

Cells of medium size, about 6 times longer than wide, spindle-shaped, lateral margins almost perfectly straight from near the center to the narrowly rounded poles. Cell wall smooth, colorless. Chloroplast with about 5-8 continuous longitudinal ridges and 5-9 pyrenoids in a median row.

### 158) *C. rectimarginatum* Scott *and,* Prescott, v. *andamense* Prasad and Misra (Pl. 18, Fig. 6)

Prasad and Misra, 1992, p 116, pl 15, f 16

Cells large, longer than broad, spindle shaped, lateral margins almost straight and coverging from centre to narrowly rounded apices; cell wall smooth; chloroplast with 6 ridges, containing large pyrenoids. Long. cell 610 μ, lat. cell 50 μ, lat. apex 7.5 μ.

### 159) *C. sigmoideum* Lagerh. *et.* Nordst. (Pl. 18, Fig. 7)

Prasad and Misra, 1992, p 118, pl 15, f 4

Cells of medium size, about 7 times longer than broad, faintly sigmoid, middle portion more or less straight and gradually attenuated towards moderately recurved and obtusely rounded apices; cell wall smooth; Chloroplast with ridges and 7-8 pyrenoids arranged in somewhat irregular median row. Long. cell 210 µ, lat. cell 32.5 µ, lat. apex 5 µ.

### 160) *C. cynthia* De Not. (Pl. 18, Fig. 8)

Prasad and Misra, 1992, p 103, pl 16, f 25

Cells rather small and stout, slightly curved, outer margin 125-130 degrees of arc, inner margin parallel to outer near apices but more or less straight in the middle, cell gradually attenuated towards obtusely rounded apices; cell wall finely striate; chloroplast with 6 pyrenoids, arranged in row. Long. cell 95 µ, lat cell 15 µ, lat. apex 2.5 µ.

## Genus: *Penium* Brebisson

### 161) *Penium spinospermum* Josh. (Pl. 19, Fig. 1)

Prasad and Misra, 1992, p 95, pl 15, f 10

Cells 2.5 times longer than broad, unconstricted, very slightly attenuated towards rounded apices; cell wall smooth; each chloroplast with three ridges and one pyrenoid. Long. cell 25.5 µ, lat. cell 10.5 µ.

## Genus: *Pleurotaenium* Naegeli

### 162) *Pleurotaenium ehrenbergii* (Breb.) De Bary (Pl. 19, Fig. 2)

Prasad and Misra, 1992, p 124, pl 18, f 10

Cells fairly large, longer than broad, sub-cylindrical, slightly constricted at the base; semicells cylindrical, gently attenuated from base towards apex; basal inflation small with one undulation; apex with a ring of tubercles; cell wall minutely punctate. Long. cell 410 µ, lat. cell 25 µ, lat. apex 15 µ.

## Genus: *Micrasterias* C.A. Agardh

### 163) *Micrasterias foliacea* Bail. (Pl. 19, Fig. 3)

Prasad and Misra, 1992, p 141, pl 20, f 6

Cells rather small, rectangular in outline, deeply constricted, sinus narrowly linear; semicells five lobed, polar lobe exserted, basal part of polar lobes with sub parallel sides, upper part greatly expanded and anvil shaped with an excavation in the median portion, base of excavation exhibits two asymmetrically produced spines of unequal length, polar and lateral angles uncinate, lateral lobe asymmetrical, superior lobes divergent, inferior horizontally disposed, incision simple and subacuminate, the ultimate loblets with truncate emarginate apices; cell wall smooth. Long. cell 57.5 µ, lat. cell 72.5 µ, lat. isthmus 15 µ.

## 164) *M. mahabuleshwarensis* Hobson (Pl. 19, Fig. 4)

Prasad and Misra, 1992, p 142, pl 20, f 7

Cells of medium size, 1-5 times longer than broad, constriction deep, sinus open with acuminate extremity; semicells three lobed, with symmetry in three planes, incision between lobes wide, polar lobe large with subquadrate lower half and dialated upper half producing prominent diverging, denticulate processes and exhibiting a pair of small, accessory assymetrical denticulate processes in front and back near slightly concave base, margin with small and acute spines, lateral lobes with wide, deep and acute angled incision, divided into two attenuated and denticulate processes. Long. cell 122.5 μ, lat. cell 90 μ, lat. isthmus 22.5 μ.

## 165) *M. pinnatifida* (Kuetz.) Ralfs. (Pl. 19, Fig. 5)

Prasad and Misra, 1992, p 143, pl 20, f 4

Cells small, slightly broader than long, deeply constricted, sinus linear but slightly open outwards; semicells 3 lobed, interlobular incision deep and broadly rounded, lateral lobes horizontal, semifusiform with minutely bifid, apices exihibiting acuminate ends, polar lobe with basal portion subrectangular and apical portion with extremities like lateral lobes but relatively shorter in length; cell wall minutely punctuate. Long cell 45 μ, lat cell 55 μ, lat isthmus 12.5 μ.

## 166) *M. radians* Turner (Pl. 19, Fig. 6)

Prasad and Misra, 1992, p 144, pl 20, f 2

Cells of medium size, subcircular, very deeply constricted, sinus with apical portion linear and outer open; semicells 5 lobed with deep, radial and widely open polar lobe with subparallel sides showing retusely emarginate and somewhat expanded apex with furcate acuminate extremity, each lateral lobe divided into two lobules by incision as deep as between polar and lateral lobes; lobules with furcate acuminate extremities; cell wall smooth. Long. cell 107.5 μ, lat. cell 97.5 μ, lat. isthmus 15 μ.

## 167) *M. zeylanica* Fristch (Pl. 19, Fig. 7)

Prasad and Misra, 1992, p 147, pl 20, f 5

Cells almost as long as broad, deeply constricted, sinus narrowly linear and slightly open outwards; semicells scarely 5 lobed, incision between the polar and lateral lobes, deep and slightly open, a incision between the lateral lobes rather shallow, polar lobe broadly cuneate with faintly retuse outer margin, polar and lateral angles somewhat acuminate, each furnished with a small, inclined and subacute spine; cell wall smooth. Long. cell 50 μ, lat. cell 52 μ, lat. isthmus 10 μ.

## Genus: *Euastrum* Ehrenberg

### 168) *Euastrum ansatum* Ralfs v. *pyxidatum* Delp. (Pl. 20, Fig. 1)

Prasad and Misra, 1992, p 134, pl 19, f 1

Cells small, twice as long as broad, deeply constricted, sinus narrowly linear; semicells with large undulations above the basal angles; cell wall with punctations arranged in distinct vertical series, apices subquadrate with rounded angles. Long. cell 70 µ, lat. cell 37.5 µ, lat. isthmus 7. 5 µ.

### 169) *E. bidentatum* Nag. (Pl. 20, Fig. 2)

Rai and Misra, 2008, p 49, pl 2, f 4

Cells 45 µ long, 30 µ broad, deeply constricted, sinus slightly dialated at the extremity; isthmus narrow 6 µ wide; semicells three lobed, polar lobe with deep median incision, apical angles with short spines; lateral lobes bilated, semicells with 5 protuberances, one large just above the isthmus, one on each lateral lobe and one on each side of apical notch in the polar lobe.

### 170) *E. coralloides* Josh. v. *trigibberum* Lagerh. (Pl. 20, Fig. 3)

Prasad and Misra, 1992, p134, pl 19, f 4

Cells small with deep constriction; semicells three lobed, polar lobe short and broad with deep median incision, apical angles furnished with short spines, lateral lobe bilobulate, each lobule truncate, emarginate with a small spine in the furrow; each semicell with 5 tumours, two small and lateral in position, one rather large just above the sinus and one on each side of apical notch in polar lobe. Long. cell 37.5 µ, lat. cell 25 µ, lat.isthmus 5 µ.

### 171) *E. elegans* (Breb) Kutz. (Pl. 20, Fig. 4)

Rai *et al*, 2008 p 60, pl 1, f 3

Cells very small, longer than broad, deeply constricted, sinus narrowly linear; semicells ovate-pyramidate, basal angles subrectangular, retuse emarginate, above with the sides concave, superior angles apiculate, apex strongly convex with a deep and narrow median incision; semicells with a triverrucose or a granulate protuberance in the centre above the isthmus, the rest of the cell-wall generally smooth. Side view of semicell ovate, with a truncately-rounded apex, and an emarginated protuberance on each side above the base. Length 30 µ; breadth 22 µ; isthmus 5 µ

### 172) *E. irregulare* Gonzalves *et.* Gangla (Pl. 20, Fig. 5)

Ashtekar, 1980, p 110, pl 36, f 213

Cells usually 1½ times larger than broad, 40 µ broad, 65 µ long, isthmus 15 µ broad, deeply constricted, sinus narrowly linear, open, dilated at apex, semicells pyramidate, 5 lobed, polar lobe rounded at apex with no median incision, the lower lateral lobes rounded and unequal, much larger than the upper ones; semi cells in the centre with a large granular protuberance and with smaller granulate. Protuberance within each of the lateral lobes and within the polar lobes, semicells in lateral view somewhat rectangular with a protuberance on each side near the base with an apical granule, elliptic in vertical view, with rounded poles and with three large rounded protuberance on each side, cell wall irregularly granulate, granules especially prominent on all the rounded angels; chloroplast solitary in each semicell with a single central pyrenoid.

### 173) *E. platycerum* Reinsch (Pl. 20, Fig. 6)

Rai and Misra, 2008, p 50, pl 2, f 7

Cells 42 μ long, 37.5 μ broad, deeply constricted, sinus narrow, slightly dialated and widely open out; isthmus 10 μm wide; semicells three lobed; polar lobes truncate without median constriction, broadly rounded angles with two small marginal spines; lateral lobes broadly rounded with marginal spines; semicells with a rounded central protuberance just above the isthmus.

### 174) *E. pseudotuddalense* Messik (Pl. 20, Fig. 7)

Sahin and Akar, 2007, p 1829, Fig. 3.11

Cells are longer than wide, 20 μ long, 15 μ wide and isthmus 7.5 μ. Semicells trapezoid with three lobes. Basal lobe truncated bearing two marginal granules. Apical lobe trapezoid with concave sides. Apical incision obtuse angled with granules on each side. The incision between the basal lobes and apical lobe acute angled. Sinus deep and narrow.

### 175) *E. spinulosum* Delp. (Pl. 20, Fig. 8)

Prasad and Misra, (1992), p136, pl 19, f 10

Cell rather small, slightly longer than broad, deeply constricted, sinus narrow and liner; semi cells 5 lobed, lateral lobes rounded and furnished with 5-6 small and acute spines, polar lobe small, broadly truncate with a shallow median notch and angles furnished with 3 small and acute spines; cell wall granulate within the polar and lateral lobes, each semi cell with a rounded, central protuberance, consisting of two rows of relatively larger granules. Long. cell 60 μ, lat. cell 45 μ, lat. isthmus 10 μ.

### 176) *E. spinulosum* Delp. v. *inermius* Nordstedt. (Pl. 20, Fig. 9)

Prasad and Misra, (1992), p137, pl 19, f 11

This variety differs from species in broadly rounded to flattened lateral lobes and somewhat trapezoid polar lobe; cell wall with relatively bigger granules, arranged in more or less circular fashion, central protuberance with 10 big peripheral and 4 large internal granules. Long. cell 60 μ, lat. cell 52.5 μ, lat. isthmus 12.5 μ.

### 177) *E. spinulosum* Delp. v. *burmense* (West *et* west) Krieger (Pl. 20, Fig. 10)

Prasad and Misra, (1992), p 137, pl 19, f 8

Cells small, a little longer than broad, constriction deep, sinus narrow and open; each semicell 5 lobed with rounded angles; lateral lobe rotundo- truncate, polar lobe broadly truncate with faint median notch, lateral and polar lobes furnished with 5-6 unevenly disposed, prominent and acute spines; cell wall with short spines within the polar and lateral lobes and one large median and two relatively small lateral protuberances above the isthmus. Long cell 52.2 µ, lat cell 45 µ, lat. isthmus 12.5 µ.

## Genus: *Cosmarium* Corda ex Ralfs

### 178) *Cosmarium angulosum* Breb. v. *concinnum* (Rab.) W.*et* G.S.West (Pl. 21, Fig. 1)

Prasad and Misra, 1992, p 152, pl 21, f 8

Cell very small, a little longer than broad, deeply constricted, sinus narrow and linear; semicells hexagonal with sharp angles and parallel sides, apex narrow and slightly retuse; cell wall smooth. Long. cell 12.5 µ, lat. cell 10 µ, lat. isthmus 2.5 µ.

### 179) *C. auriculatum* Reinsch. (Pl. 21, Fig. 2)

Prasad and Misra, 1992, p 153, pl 22, f 14

Cells small, constriction not deep, sinus open outwards with rounded apex; semicells sub-elliptical, sides 5 undulate with sharp and pointed ridges, apex narrow and straight; cell wall with punctuations arranged in transverse series. Long. cell 45 µ, lat. cell 47.5 µ, lat. isthmus 22.5 µ.

### 180) *C. awadhense* Prasad and Mehrotra (Pl. 21, Fig. 3)

Misra and Srivastava, 2003, p 87, pl 1, f 21

Cells slightly longer than broad, constriction deep; sinus narrowly linear, semicell sub-semicircular, side with crenation, apex truncate with more or less straight margin, cell wall smooth, each semicell with single chloroplast. Long cell 30 µ long, lat. cell 28.5 µ, isthmus 7.5 µ.

### 181) *C. biretum* Breb. (Pl. 21, Fig. 4)

Prasad and Misra, 1992, p 154, pl 23, f 19

Cells rather small, a little longer than broad, very deeply constricted, sinus narrowly linear with slightly dilated extremity; semicells subrectangular with somewhat convex sides and apex; cell wall with granules arranged in distinct curved vertical series; each semicell with an axile chloroplast and two pyrenoid. Long. cell 60 µ, lat. cell 52.5 µ, lat. isthmus 17.5 µ.

### 182) *C. calcareum* Witter. (Pl. 21, Fig. 5)

Prasad and Misra, 1992, p 156, pl 24, f 18

Cells small, almost as long as broad, deeply constricted, sinus narrowly linear; semicells somewhat trapezoid, basal angles subrectangular, sides 4 crenate, apex truncate with 6-7 faint and minute crenations; cell wall with minute granules and a small granulate tumour in the centre; top view narrowly elliptic. Long. cell 20 μ, lat. cell 15 μ, lat. isthmus 5 μ.

### 183) *C. cucurbitinum* (Biss) Lutkem. v. *subpolymorphum* (Nordst.) Lutkm. (Pl. 21, Fig. 6)

Misra, Mehrotra, Jai Prakash, Srivastava, and Kishore, 2000, p 72, pl 2, f 11, 16

Cells of medium size, about twice as long as broad, moderately constricted, isthmus broad, semi cells sub-elliptical with slightly narrowed and rounded apex, cell wall minutely punctate, chloroplast with two pyrenoids in each semicell, long cell 50-85 μ, breadth 36-42 μ, breadth at isthmus 13.5-38 μ.

### 184) *C.decoratum* W. and G.S. West (Pl. 21, Fig. 7)

Misra and Srivastava, 2003, p 87, pl 1, f 27

Cells longer than broad, deeply constricted, sinus narrowly linear and bilipped semicells; semielliptic, apex flattened truncate with rounded angles margin deeply crenate, crenations emarginate, semicells exhibit two chloroplast, cell wall smooth. Long cell 51 μ, lat. cell 36 μ; isthmus 13-14 μ.

### 185) *C. dubium* Borge (Pl. 21, Fig. 8)

Scott and Prescott, 1961, p 58, pl 32, f 3

Cells longer than broad, up to 1½ times as long as broad, deeply constricted, sinus linear, narrow, not dialated at the apex; semicells truncate-pyramidate, with 4 undulations on each side, apices retuse, in side view elliptic-oblong, vertical view elliptic, with inflations on each side; long cells 33 μ, lat. cell 22 μ, Isthmus 7 μ, broad; cell wall smooth; chloroplast one; axile with a central pyrenoid.

### 186) *C. forceps* Bruehl *et* Biswas (Pl. 5.21, Fig. 9)

Prasad and Misra, 1992, p 158, pl 21, f 1

Cells of medium size almost circular, moderately constricted, isthmus broad and sinus open outwards; semicells sub semicircular with subacuminate basal angles and rounded apex; cell wall smooth; Chloroplast with folded margins, disposed in relation to two large pyrenoids. Long.cell 57.5 μ, lat. cell 50 μ, lat. isthmus 40 μ.

### 187) *C. formosulum* Hoffman. (Pl. 21, Fig. 10)

Misra and Srivastava, 2003, p 87, pl 1, f 19

Cells slightly longer than wide, deeply constricted, sinus linear, semi cells are broadly ovate, margin crenate, cell wall smooth chloroplast axile. Long cell 38 μ, lat. cell 36μ, isthmus 8.5-9μ.

### 188) *C. furcatospermum* W. *et* G.S. West v. *maius* Prasad and Misra (Pl. 21, Fig. 11)

Prasad and Misra, 1992, p 159, pl 24, f 24

Cells rather small, slightly longer than broad, deeply constricted, sinus linear; semicells truncate, sub semi-circular, lateral margins crenately granulate, apex broadly truncate and 5-6 undulate with 3 series of small granules within the whole margin, central region smooth; top view oblong elliptic; each semicell with an axile chloroplast containing a large pyrenoid in the centre. Long.cell 30 μ, lat. cell 25 μ, lat. isthmus 10 μ.

### 189) *C. granatum* Breb. (Pl. 22, Fig. 1)

Prasad and Misra, 1992, p 160, pl 21, f 20

Cells small, slightly longer than broad, sub-rhomboid to elliptic, deeply constricted, sinus linear with a dilated extremity; semicells truncate, pyramidate, basal angles rounded, apical angles obtuse, sides straight or slightly convex, apex narrowly truncate with faintly retuse margin, cell wall finely punctate; chloroplast axile with single pyrenoid. Long. cell 25 μ, lat. cell 17.5 μ, lat. isthmus 5 μ.

### 190) *C. hammeri* Reinsch v. *homalodermum* (Nordst.) W.*et* G.S.West (Pl. 22, Fig. 2)

Prasad and Misra, 1992, p 161, pl 21, f 4

Cells of medium size, about 1.5 times longer than broad, sub hexagonal, deeply constricted, sinus narrowly linear with slightly dilated apex and opening outwards; semi cells truncate pyrimidate with rounded basal angles and faintly retuse sides converging to broadly truncate apex; cell wall thick and distinctly punctate; Chloroplast axile with one pyrenoid. Long cell 55 μ, lat. cell 32.5 μ, isthmus 10 μ

### 191) *C. hammeri* Reinsch v. *schmidlei* Gronblad *et* Scott. (Pl. 22, Fig. 3)

Prasad and Misra, 1992, p 162, pl 21, f 18

Cells small, about 1.5 times longer than broad, sub hexagonal, deeply constricted, sinus narrowly linear with dilated apex; semicells with upper part of lateral margins converging and retuse, apices faintly retuse with rounded angles; cell wall thick and smooth. Long cell 40 μ, lat.cell 25 μ, lat. isthmus 7.5 μ.

### 192) *C. impressulum* Elfv. v. *crenulatum* (Naeg.) Krieger *et* Gerloff (Pl. 22, Fig. 4)

Prasad and Misra, 1992, p 162, pl 21, f 6

Cells very small, slightly longer than broad, deeply constricted, sinus narrowly linear; semicells vertically subquadrate with triundulate margin, undulations sub acute, apex narrow and retuse with rounded angles; cell wall smooth; top view broadly elliptic; chloroplast parietal with one pyrenoid in each semicell. Long.cell 30 μ, lat. cell 25 μ, lat. isthmus 5 μ.

### 193) *C. indentatum* Gronbl. v. *ellipticum* Scott and Gronbl. (Pl. 22, Fig. 5)

Scott and Prescott 1961, p 59, pl 27, f 7

Cell small, slightly longer than broad, deeply constricted, sinus narrow towards the apex, and open outwards, semicells broad, elliptic, apex usually somewhat flattened, circular in side view, elliptic in vertical view; cells 37 μ long; 25.5 μ broad; 5.5 μ isthmus. Cell wall smooth, colourless; chloroplast exile; with a pyrenoid in each semicell.

### 194) *C. lundellii* Delp. (Pl. 22, Fig. 6)

Prasad and Misra, 1992, p 163, pl 22, f 5

Cells of medium size, a little longer than broad, deeply constricted, sinus linear and open dialated at the apex, semicells sub-semicircular to sub-pyramidate, apices truncated, cell wall punctate, semi cells 31 μ long, 21 μ broad, isthmus 6 μ. Dimensions of the present alga are smaller than the type cell.

### 195) *C. lundellii* Delp. v. *corruptum* (Turn) W. et West (Pl. 22, Fig. 7)

Prasad and Misra, 1992, p 163, pl 22, f 2

Cells longer than broad, deeply constricted, sinus linear, narrow; semicells sub-circular, with rounded basal angles, truncated at apex; lateral view of semicells subcircular ventral view elliptic, cell wall punctate, cells 55.5 μ long; 55.5 μ broad, isthmus 22.5 μ. Cell wall smooth.

### 196) *C. lundellii* Delp. v. *ellipticum* West and West. (Pl. 22, Fig. 8)

Prasad and Misra, 1992, p 164, pl 24, f 10

Cells longer than broad, deeply constricted, sinus linear, narrow; semicells sub-circular, truncated at apex, vertical view elliptic; cell wall punctuate, cells 66-74 μ long; 53-56 μ broad; isthmus 18-22 μ. Cell wall smooth.

**197) *C. lundellii* Delp. v. *subellipticum* Prasad et Mehrotra (Pl. 22, Fig. 9)**

Prasad and Misra, 1992, p 165, pl 22, f 9

Cells of medium size, a little longer than broad, deeply constricted, sinus linear and open; semicells sub semicircular to sub pyramidate, basal angles rounded; cell wall coarsely punctuate. Semicells subeliptic with rounded to subtruncate apices. Long. cell 50 µ, lat. cell 37.5 µ, lat. isthmus 12.5 µ.

**198) *C. margaritatum* (Lund.) Roy *et.* Biss. (Pl. 22, Fig. 10)**

Prasad and Misra, 1992, p 165, pl 23, f 3

Cells rather large, longer than broad, deeply constricted; semicells subrectangular, apex and sides slightly convex, basal and apical angles broadly rounded, cell wall uniformly granulate and with punctae between them; chloroplast axile with two pyrenoids in each semicell. Long. cell 65 µ, lat. cell 52.5 µ, lat. isthmus 17.5 µ.

**199) *C. miscellum* Skuja (Pl. 22, Fig. 11)**

Misra and Srivastava, 2003, p 87, pl 2, f 31

Cells large, slightly longer than broad, sides converging upwards, form a broad base to flattened apex, apical angles sub-acute, isthmus bilipped, semi-cells compressed, semi-circular with crenulations, 2 axile chloroplast in each semi-cell, cell wall smooth. Long cells 42-59 µ, lat. cells 37-53 µ, isthmus 11-14 µ.

**200) *C. moniliforme* (Turnp.) Ralfs. v. *lamneticum* W. *et* G.S. West (Pl. 23, Fig. 1)**

Prasad and Misra, 1992, p 167, pl 21, f 14

Cells commonly seen attached in pair of 2 or 4, apices slightly angular, isthmus broad, sinus abtuse; cell wall smooth; each semicell with an axile chloroplast and one pyrenoid. Long.cell 30 µ, lat. cell 20 µ, Lat. isthmus 7.5 µ.

**201) *C. nitidulum* De Not. (Pl. 23, Fig. 2)**

Misra and Srivastava, 2003, p 89, pl 1, f 4, 7

Cells small, slightly longer than broad, deeply constricted, sinus narrowly linear, semi cell, truncate, sub-semicircular, tapering evenly from rounded basal angles to flattened apex with straight margin, cell wall minutely punctate, chloroplast with one pyrenoid, long cells 28-30 µ, lat. cells 23-24 µ, isthmus 6-7 µ.

**202) *C. norimbergense* Reinsch v. *depressum* West and West. (Pl. 23, Fig. 3)**

Misra and Srivastava, 2003, p 89, pl 1, f 2

Cells very small, about 1½ times as long as wide, sinus nearly open, semi-cells sub quadrate sub-rectangular, semi-cell have protuberance on both side, cell wall smooth. Long cells 18-20 μ, lat. cells 11-14 μ, isthmus 5.5-6μ, chloroplast axile with one pyrenoid per cell.

### 203) *C. norimbergense* Reinsch v. *elongatum* (W. *et* G.S. West) Krieger *et* Gerloff (Pl. 23, Fig. 4)

Prasad and Misra, 1992, p 169, pl. 21, f 12

Cells very small, about 1.4 times longer than broad, deeply constricted, sinus narrowly linear; semicells sub-rectangular, sides with a subacute undulations just above the rounded basal angles and slightly retuse in upper half, apex moderately narrowed with rounded angles and more or less straight margin; cell wall smooth; chloroplast axile with one pyrenoid in each semicell. Long. cell 17.5 μ, lat. cell 12.5 μ, lat.isthmus 2.5 μ.

### 204) *C. nymannianum* Grun (Pl. 23, Fig. 5)

Kouwets, 1987, p 226, pl 12, f 8

Cells medium sized, about 11/5 times as long as broad, deeply constricted, sinus narrowly linear, open, not dilated at the apex, semi-cells trapezoid, with basal and apical angles rounded, apices and side concave, semicells in lateral view elliptic ovate, vertical view elliptic; cell wall smooth, cells 25μ long; 18.5μ lat. cell; isthmus 4.5 μ. chloroplast with central pyrenoid.

### 205) *C. obsoletum* (Hantzsch) Reinsch. v. *sitvense* (Gutw.) Krieger (Pl. 23, Fig. 6)

Prasad and Misra, 1992, p 171, pl 22, f 13

Cells of medium size, transversly elliptic, almost as long as broad, deeply constricted, sinus narrow with dialated apex and slightly open outwards; semicells depressed, semicircular, apex somewhat flattened; cell wall distinctly punctate. Long.cell 55 μ, lat. Cell 70 μ, lat.isthmus 30 μ.

### 206) *C. obtusatum* Sch. (Pl. 23, Fig. 7)

Suseela and Toppo, 2009, p 968, pl 1, f 5

Cells small, deeply constricted, sinus narrowly linear; cell wall smooth; chloroplast axile with one pyrenoid in each semicell. Cells 27.5 μ in diameter, 37.5 μ long, isthmus 10 μ. long

### 207) *C. pachydermum* Lundell v. *aethiopicum* W. and G.S. West. (Pl. 23, Fig. 8)

Misra and Srivastava, 2003, p 89, pl.1, f 6

Cells longer than wide, broadly elliptic, deeply constricted, lower part of sides somewhat straight, basal angle broadly rounded, cell wall thin, finely scorbiculate and minutely punctate between suborbicuation. Long cells 150 µ, lat. cells 119-120 µ, isthmus 60 µ.

### 208) *C. pardalis* Cohn (Pl. 23, Fig. 9)

Scott and Prescott, 1961, p 64, pl 29, f 1

Cells longer than broad, about 1½ times as long as broad, deeply constricted, sinus narrowly linear, slightly dilated at the apex; semi cells semi-circular, ventral view elliptic; cells 62.9 µ long; lat. cells.55.5 µ broad; isthmus 18.5 µ; cells wall punctate; chloroplast with a central pyrenoiad.

### 209) *C. polygonum* (Naeg.) Arch. (Pl. 23, Fig. 10)

Prasad and Misra, 1992, p 171, pl 21, f 11

Cells small, slightly longer than broad, deeply constricted, sinus narrow; semicells broadly hexagonal, angles rounded, lateral angles slightly pronounced, sides and apex faintly retuse, cell wall smooth. Long.cell 17.5 µ, lat.cell 15 µ, lat.isthmus 2.5 µ.

### 210) *C. portianum* Arc. v. *nephroideum* Witter (Pl. 23, Fig. 11)

Prasad and Misra, 1992, p 172, pl 23, f 13

Cells small, about 1.3 times longer than broad, deeply constricted, sinus gradually opening from a rounded extremity, isthmus slightly elongated; semicells subreniform and granulate, granules rounded and evenly disposed in indistinct vertical series, about 24-29 visible at margin of each semicell; semicell with an axile chloroplast and pyrenoid. Long.cell 22.5 µ, lat.cell 17.5 µ, lat.isthmus 7.5 µ.

### 211) *C. pseudobroomei* Wolle (Pl. 24, Fig. 1)

Prasad and Misra, 1992, p 173, pl 23, f 11

Cells small, as long as broad very deeply constricted, sinus narrowly linear with dialated extremity; semicells ablong-rectangular with rounded angles and slightly convex sides and apex; cell wall with small and solid granules arranged in somewhat curved horizontal series; chloroplast axile with two pyrenoids in each semicell. Long.cell 40 µ, lat.cell 37.5 µ, lat.isthmus 10 µ.

### 212) *C. pseudoconnatum* Nordst. (Pl. 24, Fig. 2)

Prasad and Misra, 1992, p 174, pl 22, f 7

Cells of medium size, about 1.2 times longer than broad, slightly constricted, sinus broad and shallow; semicells semielliptic with circular outline; cell wall finely punctate; top view subcircular; each semicell with two axile chloroplasts, each containing one pyrenoid. Long.cell 52.5 µ, lat. cell 50 µ, lat. isthmus 37.5 µ.

## 213) *C. pseudogranatum* Nordst v. *rotundatum* (Krieg.) Messik. (Pl. 24, Fig. 3)

Prasad and Misra, 1992, p 175, pl 21, f 22

Cells small; semicells broadly truncate exhibiting rather prominently convex sides and truncate rounded apex; cell wall minutely punctate. Long. cell 22.5 μ, lat. cell 17.5 μ, lat. isthmus 5 μ.

## 214) *C. punctulatum* Breb. (Pl. 24, Fig. 4)

Prasad and Misra, 1992, p 176, pl 23, f 21

Cells small, deeply constricted, sinus narrowly linear with a dilated extremity; semicells oblong trapeziform, basal and apical angles rounded, sides convex and slightly converging upwards, apex broadly truncate; cell wall with fine granules. Long.cell 32.5 μ, lat.cell 27.5 μ, lat.isthmus 10 μ.

## 215) *C. punctulatum* Breb. v. *subpunctulatum* (Nordst.) Borge forma Scott et Prescott (Pl. 24, Fig. 5)

Prasad and Misra, 1992, p 176, pl 23, f 18

Cells small, a little longer than broad, very deeply constricted sinus with slightly dilated extremity; semi cells subquadrangular trapizeform, basal angles rounded, sides convex and converging upwards to broadly truncate apices with straight margins and sharp angles; cell wall granulate, granules small and of uniform size, disposed evenly towards the periphery, central region with 3 transverse series of slightly bigger granules. Long.cell 25 μ, lat.cell 20 μ, lat. isthmus 5 μ.

## 216) *C. pyramidatum* Breb. (Pl. 24, Fig. 6)

Prasad and Misra, 1992, p 177, pl 22, f 18

Cells of medium size, longer than broad, about 1½ times as long as broad, deeply constricted, sinus linear, open, slightly dilated at the apex; semi-cells pyramidate, basal angles rounded, sides slightly convex to straight, apices truncated, retuse, apical angles rounded, Cell wall scrobiculate, cells 44 μ long, 28μ broad, isthmus 7.5μ broad; cell wall distincly punctate, punctae sparsely arranged, chloroplast axile.

## 217) *C. quinarium* Lund. (Pl. 24, Fig. 7)

Prasad and Misra, 1992, p 179, pl 24, f 17

Cells rather small, sub hexagonal, slightly longer than broad, deeply constricted, sinus linear with slightly dilated apex and opens outwards; semi cells broadly truncate pyramidate with rounded basal and sub acute apical angles, lateral margins slightly convex, apex truncate and somewhat flattened; margin of semi cell with 14-17 acute granules followed by series of rounded granules within the outline, centre of each semicell with 2 transverse series of large granules, 3 in upper and 2 in lower series and 2 punctations between them; chloroplast axile with 2 pyrenoids. Long.cell 37.5 μ, lat. cell 27.5 μ, lat. isthmus 7.5 μ.

### 218) *C. radiosum* Wolle (Pl. 24, Fig. 8)

Prasad and Misra, 1992, p 179, pl 24, f 12

Cell of medium size, about 1.5 times longer than broad, deeply constricted, sinus narrowly linear with slightly dialated extremity; semicells sub semicircular, sides convex, apex subtruncately rounded; cell wall finely granulate, granules in radiating and concentric series towards periphery; chloroplast axile with two pyrenoids in each semicell. Long.cell 40 μ, lat. cell 30 μ, lat. isthmus 12.5 μ.

### 219) *C. regnellii* Wille (Pl. 24, Fig. 9)

Prasad and Misra, 1992, p 180, pl 24, f 25

Cells very small, a little longer than broad, deeply constricted, sinus narrow with slightly dilated extremity; semicells sub-hexagonal, basal angles more or less sub-rectangular, sides parallel, upper angles broad and oblique, apex truncate and straight; cell wall smooth; each semicell with an axile Chloroplast and one pyrenoid. Long.cell 12.5 μ, lat. cell 10 μ, lat. Isthmus 2.5 μ.

### 220) *C. reniforme* (Ralfs) Arch. (Pl. 24, Fig. 10)

Prasad and Misra, 1992, p 181, pl 23, f 8

Cells of medium size, slightly longer than broad, constriction deep, sinus narrow and linear than widely dialated extremity; semicells reniform; cell wall granulate, granules fairly regular, horizontal and indistinct vertical series; chloroplast axile with two pyrenoids in each semicell. Long.cell 55 μ, lat. cell 45 μ, lat. isthmus 15 μ.

### 221) *C. retusiforme* (Wolle) Gutw. (Pl. 24, Fig. 11)

West and West 1905, p180, pl 62, f 17, 18

Cells of medium size, longer than broad, deeply constricted, sinus linear, semi cells truncate, pyramidate, cell wall smooth, long cell 29.6 μ; lat cell 22.2 μ; isthmus 5.5 μ, each semicell with one pyrenoid.

### 222) *C. sexnotatum* Gutw. (Pl. 25, Fig. 1)

Prasad and Misra, 1992, p 182, pl 24, f 16

Cell small, sinus narrow and linear with slightly dialated apex; semicells subquadrangular with rounded angles, sides with three broad undulations, apex truncate with straight or faintly undulate margin; cell wall furnished with 2-3 curved and horizontal series of granules within the margin, top view elliptic;

Chloroplast axile with a central pyrenoid in each semicell. Long.cell 20 μ, lat. cell 15 μ, lat.isthmus 5 μ.

## 223) *C. speciosum* Lund. (Pl. 25, Fig. 2)

Prasad and Misra, 1992, p 183, pl 24, f 14

Cells of medium size, about 1.5 times longer than broad; moderately constricted, sinus narrowly linear; semicells subrectangular or subpyramidate with rounded angles, sides slightly convex and very gradually attenuated upwards to broadly truncate apex, margin with four apical and seven lateral crenations; cell wall granulate, granules in regular, radial and concentric series, each showing 3-4 granules, space across the base and just above the isthmus exhibits 5-6 vertical series of 6-7 granules; chloroplast axile with two pyrenoids in each semicell. Long. cell 45 μ, lat. cell 32.5 μ, lat. isthmus 10 μ.

## 224) *C. subalatum* W. *et* G.S.West (Pl. 25, Fig. 3)

Prasad and Misra, 1992, p 185, pl 24, f 22

Cells small, slightly longer than broad, deeply constricted, sinus narrowly linear; semicells widely truncate to pyramidate, sides tricrinate, angles rounded, apex with two small crenations, crenations bigranulate, central tumour rounded with seven granules arranged in circular fashion, chloroplast axile with one pyrenoid in each semicell. Long. cell 25 μ, lat. cell 20 μ, lat. isthmus 5 μ.

## 225) *C. subcrenatum* Hantzsch (Pl. 25, Fig. 4)

Prasad and Misra, 1992, p 185, pl 24, f 25

Cells small, slightly longer than broad, deeply constricted, sinus narrowly linear; semicells sub-semicircular with broadly rounded or somewhat flattened apex; margin crenate-granulate and followed by 2-3 radial and concentric series of granules, centre of semicell with a broad and flattened tumour consisting of irregularly arranged, slightly bigger granules; top view elliptic with a broad tumour on each side. Long.cell 32.5 μ, lat. cell 25 μ, lat. isthmus 7.5 μ.

## 226) *C. subimpressulum* Borg (Pl. 25, Fig. 5)

Tiffany and Britton, 1952, p186, pl 54, f 588

Cells small, longer than broad, deeply constricted, sinus linear, narrow; cell wall smooth, semicells transversely rectangular in lower part, pyramidate-truncate above, with 4 crenulations on each side, lateral view of semicells ovate, apices truncate, tumid in the middle portion on both sides, cells 25.9 μ; lat. cells 18.5 μ; isthmus 5.5 μ.

## 227) *C. turgidum* Breb. v. *minus* (Reinsch) Schmidle (Pl. 25, Fig. 6)

Prasad and Misra, 1992, p 187, pl 24, f 1

Cells large, longer than broad, faintly constricted; semicells widely ovate with broad base and rounded apex; cell wall sparsely punctate; chloroplast parietal with two pyrenoids in each semicell. Long.cell 105 μ, lat. cell 72.5 μ, lat. isthmus 65 μ.

**228) *C. turgidum* Breb. v. *subrotundatum* West (Pl. 25, Fig. 7)**

Prasad and Misra, 1992, p 187, pl 24, f 7

Cells large, longer than broad with shallow constriction, semicells slightly inflated and subcircular; cell wall distinctly and sparsely punctate. Long.cell 102.5 μ, lat.cell 62.5 μ, lat.isthmus 57.5 μ.

**229) *C. turgidum* (Breb.) Ralfs v. *westii* Bernard Krieger and Gerloff. (Pl. 25, Fig. 8)**

Misra and Srivastava, 2003, p 90, pl 1, f 25

Cells about twice as long as broad constriction shallow with somewhat rounded angles sides faintly convex and truncately rounded, widely ovate, cell wall with minute punctation, arranged in indistinct transverse series. Long cells, 140μ, lat. cells 98μ, isthmus 90μ.

**230) *C. undulatum* Corda ex Ralfs v. *minutum* Wittr. Forma Suxena *et* Venkateswarlu (Pl. 25, Fig. 9)**

Prasad and Misra, 1992, p 188, pl 21, f 15

Cells small, quadrate-elliptic, slightly longer than broad, deeply constricted, sinus narrow with dilated extremity; semicells sub-semicircular with 7-8 shallow undulations, apex slightly retuse; cell wall smooth; top view elliptic; chloroplast axile with one pyrenoid in each semicell. Long. cell 22.5 μ, lat. cell 17.5 μ, lat. isthmus 2.5 μ.

**231) *C. variolatum* v. *rotundatum* (Krieger) Messik (Pl. 25, Fig. 10)**

Ashtekar, 1980, p 138, pl 41, f 271

Cells longer than broad, about 1 ½ times as long as broad, deeply constricted, sinus narrowly linear, slightly dialated at the apex, semicells triangular to subpyramidate, semicircular, vertical view elliptic; cells 29.6 μ long; lat. cells 20 μ broad; isthmus 5.5 μ. Cell wall punctuate; chloroplast with cental pyrenoid.

## Genus: *Arthrodesmus* Ehrenberg

**232) *Arthrodesmus convergens* Ehrneb. (Pl. 26, Fig. 1)**

Prasad and Misra, 1992, p 190, pl 25, f 5

Cells medium size, about 1.4-1.5 times longer than broad, median constriction deep, sinus opening widely outwards from narrowly linear extremity; semi cells more or less transversely elliptic, apical margin slightly more convex than the basal, lateral angles rounded conical and furnished with rather stout and slightly incurved spine; cell wall smooth; chloroplast axile with 1-2 pyrenoids in each semi cell. Long.cell 35 μ, lat.cell with spine 50 μ, lat.cell without spine 37.5 μ, lat.isthmus 10 μ.

### 233) *A. curvatus* Turner *f. major* Turner (Pl. 26, Fig. 2)

Prasad and Misra, 1992, p 191, pl 25, f 1

Cell of medium size, almost as long as broad, deeply constricted, sinus narrow with slightly dilated extremity and open outwards; semicells transversly oblong-elliptic with straight or faintly retuse apices, lateral angles acute with long and converging spines showing slightly recurved tips; cell wall smooth; chloroplast axile with one pyrenoid. Long. cell 45μ, lat. cell with spine 92.5 μ, lat. cell.with out spine 40 μ, lat. isthmus 12.5 μ.

## Genus: *Staurastrum* Meyen

### 234) *Staurastrum bieneanum* Rabenh. v. *ellipticum* Wille (Pl. 26, Fig. 3)

Prasad and Misra, 1992, p 194, pl 26, f 2, 5

Cells small, slightly broader than long, very deeply constricted, sinus widely open with an acuminate apex; semicells narrowly elliptic with rounded angles; cell wall punctate, punctations arranged in concentric series around the angles. Long cell 30.5 μ, lat. cell 35 μ, isthmus 10 μ.

### 235) *S. gracile* Ralfs f. Iyengar *et.* Vimala Bai (Pl. 26, Fig. 4)

Prasad and Misra, 1992, p 197, pl 25, f 14, 18

Cell small, about 2.7 times longer than broad with slight contriction in the form of an acute notch; semicell slightly broading towards the faintly convex apex, upper angles produced into more or less horizontally disposed long processes tipped with 3 minute spines and showing many concentric series of denticulations; top view triangular; chloroplast axile with one pyrenoid in each semi cell. Long.cell 17.5 μ; lat.cell 12.5 μ, isthmus 5 μ.

### 236) *S. gracile* Ralfs. v. *coronulatum* Boldt. (Pl. 26, Fig. 5)

Misra, Mehrotra, Jai Prakash, Srivastava and Kishore, 2000, p 72, pl 2, f 7, 22

Cells of medium size, about 2-3 times longer than broad (excluding the processes), depressed, constriction shallow with an acute notch; semi-cells somewhat broadening towards the slightly convex apex, apices showing undulate margins and relatively shorter and emarginate processes tipped with 2-3 minute spines and showing 4-5 concentric rows of dentations, top view triangular with dentations within the lateral margins. Long cell 29 µ, lat. cell with processes 44 µ, lat. isthmus 12 µ.

### 237) S. *gracile* Ralfs v. *kriegeri* Scott and Prescott (Pl. 26, Fig. 6)

Scott and Prescott, 1961, p 94, pl. 57, f 11

Cells small, about 2.7 times longer than broad with slight constriction in the form of an acute notch; semicells slightly broading towards faintly convex apex, upper angles produced into more or less horizontally disposed processes, processes are convergent in front view; top view triangular; chloroplast with one pyrenoid in each semicell. Long.cell 15 µ, lat.cell 20 µ, lat.isthmus 5 µ.

### 238) S. *granulosum* (Ehrenb.) Ralfs (Pl. 26, Fig. 7)

Prasad and Misra, 1992, p 197, pl 25, f 22

Cells small, almost as long as broad, deeply constricted, sinus open with subacute apex; semicells more or less obsemicircular with convex apex, lateral angle scarcely rounded and furnished with acute spines; top view triangular with slightly retuse median portion; cell wall granulate, granules minute, arranged in concentric series near the lateral angles and scattered in middle portion. Long cell 27.5 µ, lat. cell with spine 32.5 µ, lat isthmus 10 µ

### 239) S. *leave* Ralfs (Pl. 26, Fig. 8)

Misra *et al.*, 2005, p 78, f 4

Cells slightly longer than broad; cells 27.5 µ long, 25 µ broad (with processes); isthmus 7.5 µ wide.

### 240) S. *pachyrhynchum* Nordst. (Pl. 26, Fig. 9)

Prasad and Misra, 1992, p 198, pl 25, f 21

Cell rather small, almost as long as broad, deeply constricted, sinus open and acute angled; semicell sub elliptic, dorsal margin strongly convex, angles thickned, obtusely rounded and slightly produced with a faint upward tilt; top view triangular showing convex sides; cell wall smooth. Long cell 27.5 µ; lat. cell 25 µ, isthmus 10 µ wide.

### 241) S. *sexangulare* Lund v. *productum* Nordst (Pl. 26, Fig. 10)

Rai and Misra, 2008, p 53, pl 2, f 9

Cells 17.5-30 µ long (without processes) and 55 µ broad (with processes) broad; semicells with 5 large processes, each large process provided with a small process at its lower side at the base, each process terminate into 3-5 divergent spines.

### 242) *S. sexcostatum* Breb.ex Ralfs v. *productum* (W.West) G.S. West (Pl. 27, Fig. 1)

Rai *et al*, 2008, p 65, pl 1, f 22

Cell 40 µ long, 42.5 µ broad with processes, shallow constriction with an acute notch; isthmus 15 µ wide; semicells nearly hexagonal to cup shaped, dorsal margin straight to slightly convex. Processes six, stought, broad at the base, chloroplast with one pyrenoid.

### 243) *S. wildemanii* Gutw. (Pl. 27, Fig. 2)

Prasad and Misra, 1992, p 200, pl 26, f 3

Cells rather small, a little longer than broad, deeply constricted, sinus open with acuminate apex; semicells transversely subeliptical with slightly convex apex showing a long and subacute spine, ventral margin with more curvature, lateral angles somewhat truncate and furnished with a pair of diverging spines; cell wall smooth. Long. cell with spine 32.5 µ, long cell without spine 25 µ, lat. cell with spine 27.5 µ, lat.cell without spine 22.5 µ, lat isthmus 7.5 µ.

## Genus: *Xanthidium* Ehrenberg

### 244) *Xanthidium sexmamillatum* W and G.S. West (Pl. 27, Fig. 3)

Rath and Adhikari, 2005, p 57, pl. 8, f 25, pl 17, f 131

Cells fairly big, little longer than broad, deeply constricted, sinus broadly open, semicells transversly elliptic, lateral margin with 3 mamillae on each side, one apical, one subapical and one median. Six strong spines on the apices of these six mamillae on each semicell, apical and subapical spines curved upwards, the median being horizontal, forth mamillae distinctly marked with a short spine on both the lower side of the semicell, apex mostly straight, vertical view nearly rhomboidal, sides thickened slight more yelloeish, three asymmetrically disposed spines at each pole, cell wall punctate, pyrenoids two in each semicell.

## Genus: *Desmidium* C.A. Agardh

### 245) *Desmidium baileyi* (Ralfs) Nordst. v. *baileyi* f. *tetragonum* Nordst. (Pl. 27, Fig. 4)

Rai and Misra, 2008, p 54, pl 1, f 9

Cells 15 µ long, 20 µ broad, rectangular with parallel lateral margins and broad apices with deep and semi-elliptic depression, median constriction reduced into undulation, united into a straight filament without gelatinous sheath; isthmus 15 µ wide; cell wall smooth; chloroplast axile.

### 246) *D. swartzii* (Ag.) Ag. ex Ralfs (Pl. 27, Fig. 5)

Rai and Misra, 2008, p 54, pl 1, f 1 and 4

Cells 15 µ long, 25 µ broad, narrowly rectangular, moderately constricted, sinus linear but widely open out, united into spirally twisted filaments; isthmus 20 µ wide; semicells narrowly oblong, lateral margins obliquely truncate with upper angle protruded towards the apex, apex slightly staraight without depression; cell wall smooth; chloroplast axile.

## Genus: *Onychonema* Wallich

### 247) *Onychonema leave* Nordst v. *latum* West et West. (Pl. 27, Fig. 6)

Misra *et al.*, 2005, p 78, f 9

Cells wider than long, compressed, united into filament without a gelatinous sheath; cells 17.5 µ long, 27.5 µ broad; semicells broadly elliptical or subreniform, lateral apex acute with spine.

## Genus: *Sphaerozosma* Corda

### 248) *Sphaerozosma excavata* Ralfs. (Pl. 27, Fig. 7)

Smith, 1950, p 330-331,f 242 (B)

Apical processes of the semicells are short and interlocked with one another, cells are small, compressed, moderately constricted with an open sinus. Cell wall granulates; Chloroplasts are axial in each semi cell, and with single pyrenoid.

## Genus: *Triploceras* Bailey

### 249) *Triploceras gracile* Bailey (Pl. 27, Fig. 8)

Opute, 2000, p 141, pl 6, f 9

Cell about 14 times longer than broad; this is stout variety with more elaborate ornamentations, arranged in whorls about 12 of them in each semicell; each whorl has a circle of acute thickened projections or teeth; semicells tapering to a 2-lobed apex, each lobe tipped with three spines.

## Order: Tetrasporales

## Family: Palmellaceae

## Genus: *Asterococcus* Scherffel

### 250) *Asterococcus limneticus* G.M. Smith (Pl. 27, Fig. 9)

Prescott, 1951, p 86, pl 4, f 11

Cells spherical, arranged at some distance from one another in free-floating colonies of 4-16 within colourless homogenous investing mucilage. Chloroplast stellate with 4-16 lobes radiating from a central core, the lobes becoming flattened against the cell wall. Cells 20 μ in diameter; colonies 112.5 μ in diameter.

## Order: Charales

## Family: Characeae

## Tribe: Nitelleae

## Genus: *Nitella* Agardh (em. Leonh)

### 251) *Nitella bengalensis* Kundu (Pl. 28, Fig. 1)

Pal, Kundu, Sundaralingam and Venkataraman, 1962, p 67-68, f 100-104

Monoecious; stem 300 μ in diameter, whorls of 6 branchlets; sterile and fertile whorls dissimilar; sterile whorl branchlets long; 2 to 3 times; oogonia solitary, usually along with an antheridium at the second and third function of the branchlets, antheridium 300 μ in diameter, oogonium 450 μ long, 330 μ broad, spiral cells 9 convolutions; corona persistent, occasionally deciduous; 34 μ high; oospores subglobose, antheridium 220 μ in diameter.

### 252) *N. microcahateirpa* Braun (Pl. 28, Fig. 2)

Pal, Kundu, Sundaralingam and Venkataraman, 1962, p 69, f 109

Monoecious; stem moderately slender to stout, 750μ thick; upper internodes ½ to as long as the branchlets, lower internodes some-what longer; sterile branchlets 6, 2-3 time furcate; primary rays ½ as long as the entire branchlets sometimes longer; male and female gametangia at all and the same nodes, antheridia sessile; solitary, terminal, 285 μ in diameter; oogonia sessile, lateral frequently 2-4, 330μ long, 195μ wide; spiral cells showing 7-8 convolutions; coronula 70 μ high; 90 μ wide at base, persistent.

## Tribe: Chareae

## Genus: *Chara* L.

### 253) *Chara brachypus* Braun (Pl. 28, Fig. 3)

Pal, Kundu, Sundaralingam and Venkataraman, 1962, p 111, f 259-260

Monoecious, slightly incrusted, very brittle, 18 cm high; stem rather slender, upto 600 μ in diameter; internodes as long as or longer than the branchlets; cortex triplostichous, spine cells rather frequent, short, solitary, upto 60 μ long; stipulodes in a double whorl, twice as numerous as the branchlets, branchlets 9-12 in a whorl, consisting of 5-8 segments, bract-cells 6-8 on fertile nodes 4 μ long, 44 μ wide; antheridia and oogonia at the three lowest nodes, solitary at the same nodes; globule (antheridia) 300µ in diameter; nucule (oogonium) 825 µ long, 555 µ wide.

### 254) *C. braunii* Gmelin (Pl. 28, Fig. 4)

Pal, Kundu, Sundaralingam and Venkataraman, 1962, p 89, f 200-203

Monoecious, up to 37 cm high, stem moderately stout, 480 µ in diameter; internodes variable in length, usually as long as the branchlets; stem and branchlets entirely ecorticate; stipulodes in a single whorl, well-developed, as numerous as the branchlets and alternating, acute; branchlets 10, consisting of 4-6 segments, bract-cells 3-4, small, acute, antheridia and oogonia at the first, second lowest node but not at the base of the branchlet-whorls, solitary, seldom double, at the same nodes; antheridia 360 in diameter; oogonia 675 µ long, 480 µ wide.

### 255) *C. delicatula* (Agardh) A. Braun (Pl. 28, Fig. 5)

Pal, Kundu, Sundaralingam and Venkataraman, 1962, p 110-111, f 262-266

Monoecious, up to 14 cm long, sometimes incrusted; stem slender, 420 µ in diameter; internodes as long as twice as long as the branchlets; spine cell usually minute, papilliform; stipulodes in a double whorl, branchlets 7 in a whorl, usually curved, composed of 8-11 segments, bract-cells 5-7, anterior pair only developed, bracteoles somewhat longer than the oogonium; antheridia and oogonia together at the three lowest branchlet nodes, solitary, antheridium 420 µ in diameter; oogonia 975 µ long, 525 µ wide; spiral cells showing 14-15 convolution.

### 256) *C. flaccida* Braun (Pl. 28, Fig. 6)

Pal, Kundu, Sundaralingam and Venkataraman, 1962, p 95, f 220-222

Monoecious, up to 32 cm high, heavily incrusted; stem slender 392-525 µ in diameter; internodes 1-3 times the length of the branchlets; stem corticated, diplostichous; cells of the primary series more prominent than the secondry; spine cell single, acute projecting horizontally, stipulodes forming a single whorl, well developed, elongated, acute, branchlets 8-12, consisting of 4-6 segments, slender; bract-cells 4-5 at the lowest nodes, 3 at the apex, straight, thick; bracteoles similar to the anterior bract-cells; antheridia and oogonia usually at the lowest three nodes, antheridia 420 µ in diameter; oogonia up to 780 µ long, 525 µ wide.

### 257) *C. hatei* Dixit (Pl. 28, Fig. 7)

Pal, Kundu, Sundaralingam and Venkataraman, 1962, p 101, f 227-228

Plant minute; 3 cm long; trailing on the soil, monoecious; stem slender 390 µ thick; stem cortex; a few lower internodes partly ecorticate; spine cells minute, almost rudimentary; stipulodes in a single circle; long pointed, branchlets 10;

longer than the internode; segments usually five, ecorticate, bracts 5, long; 0.25-0.75 μ the length of the segments, gametangia at the lower two nodes; nucule (oogonia) 480 μ long, 300 μ broad; globule (antheridia) 240 μ in diameter.

### 258) *C. hydropitys* Reich *f. major* A. Br. (pl. 28, Fig. 8)

Pal, Kundu, Sundaralingam and Venkataraman, 1962, p 98, f 225

Monoecious; stem rather slender; plants 10 cm high; stem 390 μ thick, internodes as long as the branchlets, or slightly longer; cortex diplostichous, primary rows more prominent; branchlets 10-11, straight, with 5-6 segments, stipulodes single row, well developed; spine cells, solitary, rudimentary; bracts and bracteoles 8; gametangia produced from the $2^{nd}$ to the $5^{th}$ nodes; antheridium 225 μ in diameter, oogonium 420 μ long and 300 μ broad, ellipsoidal.

## Class: Bacillariophyceae

## Order: Bacillariales

## Sub order: Coscinodiscineae

## Family: Coscinodiscaceae

## Genus: *Cyclotella* Kuetz.

**1) *Cyclotella glomerata* Bachman (Pl. 29, Fig. 1)**

Sarode and Kamat, 1984, p 21, pl. 1, f. 10

Frustules small, in loose chains, valves discoid, rectangular in girdle view; 12.5 μ in diameter, central field smooth with 6-7 small radial striae or dots in a ring, striae 8-10 in 10 μ.

**2) *C. meneghiniana* Kuetz. (Pl. 29, Fig. 2)**

Sarode and Kamat, 1984, p 21, pl 1, f 11

Frustules discoid in valve view, rectangular in girdle view with undulate valves, 15 μ in diameter, margins strong; central field large and finely punctate, striae 8-10 in 10 μ, thick.

## Sub order: Fragilarineae

## Family: Fragilariaceae

## Genus: *Fragillaria* Lyngbye

**3) *Fragilaria brevistriata* Grun. v. *vidarbhensis* Sarode and Kamat (Pl. 29, Fig. 3)**

Sarode and Kamat, 1984, p 26, pl 1, f 18.

Frustules linear, loosely attached together to form short chains in girdle view; valves 19.7-23μ long, 3.5-4.8μ broad, linear lanceolate, strongly tumid in the middle and slightly inflated towards the ends with somewhat acutely rounded ends; pseudoraphe broad in the middle, linear lanceolate, striae 11-13 in 10μ thick.

**4) *F. intermedia* Grun. (Pl. 29, Fig. 4)**

Sarode and Kamat, 1984, p 27, pl 1, f. 21

Frustules united together to form long bands, linear, rectangular in girdle view; valves 80 μ long, 7.5 μ broad, linear with parallel margins; ends gradually tapering and rounded; axial area narrow, linear; striae 10-12 in 10 µ, coarse and distinct, absent on one side in the middle.

### 5) *F. intermedia* Grun. f. *chandrapurensis* Sarode and Kamat (Pl. 29, Fig. 5)

Sarode and Kamat, 1984, p 27, pl 1, f 22

Frustules linear, rectangular, united together to form long band in girdle view; valves 40.2-69µ long, 8.5-10µ broad, linear with parallel margins, ends abruptly tapering and rounded; pseudoraphe narrow, linear; striae 10-12 in 10 µ coarse and distinct, absent in the middle on one side.

## Genus: *Lichmophora* Agardh

### 6) *Lichmophora abbreviata* Agardh (Pl. 29, Fig. 6)

Rath and Adhikari, 2005, p 78, pl. 12, f. 71

Frustules in girdle view cuneate with strongly rounded angles, lower end attached to mucous stalk; cells forming colonies, septa projecting into the cells, valves oblanceolate with margins sub parallel towards the apex and narrowed and elongated towards the base, 42.5 µ long and 10 µ broad, pseudoraphe distinct.

## Genus: *Synedra* Ehr.

### 7) *Synedra acus* Kuetz v. *acula* (Kuetz.) V.H. (Pl. 29, Fig. 7)

Sarode and Kamat, 1984, p 31, pl. 2, f. 33

Valves 152 µ long, 5 µ broad, narrow, linear and needle like with slightly capitate ends; pseudoraphe narrow; central area without striae; striae 16-18 in 10 µ, delicate.

### 8) *S. ulna* (Nitz.)Ehr. (Pl. 29, Fig. 8)

Sarode and Kamat, 1984, p 31, pl. 2, f. 37

Valves 110 µ long, 5 µ broad, narrow, linear to linear lanceolate and tapering at the ends; ends rounded; pseudoraphe narrow, linear; central area absent or present; striae 9-11 in 10 µ, coarse.

### 9) *S. ulna* (Nitz.) Ehr. v. *notata* Kuetz. (Pl. 29, Fig. 9)

Sarode and Kamat, 1984, p 32, pl. 2, f. 42

Cells solitary and free floating or in free floating colonies or epiphytic, frustules elongated and with truncate apices in girdle view. Valves 55.6-65µ long, 8-9.5µ broad, linear with slightly narrowed and obtuse or strongly narrowed and acutely rounded ends; pseudoraphe narrow; central area absent or sometimes unilaterally widened, striae 10-11 in 10µ, coarse.

## Sub order: Eunotiineae

## Family: Eunotiaceae

## Genus: *Eunotia* Ehr.

### 10) *Eunotia arcus* Ehr. v. *uncinata* Grun. (Pl. 29, Fig. 10)

Sarode and Kamat, 1984, p 36, pl. 2, f. 48

Valves 45.5 µ long, 7.5 µ broad, arcuate, ventral side very slightly concave and dorsal side slightly convex with ends obliquely truncate and slightly constricted on the dorsal side; striae 8-9 in 10 µ.

### 11) *E. lunaris* (Ehr.) Grun. v. *subarcuata* (Naeg.) Grun. (Pl. 29, Fig. 11)

Sarode and Kamat, 1984, p 40, pl. 3, f. 63

Valves 17.5 µ long, 5 µ broad, dorsal margin arcuate, ventral margin slightly concave; ends rounded; striae 14-16 in 10 µ, fine.

### 12) *E. major* (W. Smith) Rabh. v. *indica* (Grun.)A. Berg. (Pl. 29, Fig. 12)

Sarode and Kamat, 1984, p 40, pl. 3, f. 65

Valves 42.5 µ long, 10 µ broad, sublinear, arcuate, ventral margin slightly concave and dorsal margin convex; ends constricted on dorsal side, obliquely capitate and wedge shaped; polar nodules small; striae 9-10 in 10 µ in the middle and 11-12 in 10 µ at the ends.

### 13) *E. praerupta* Ehr. (Pl. 29, Fig. 13)

Sarode and Kamat, 1984, p 44, pl. 3, f. 78

Frustules linear rectangular in girdle view; valves 35 µ long, 10 µ broad, arcuate with slightly concave ventral margin and somewhat parallel to dorsal margin; slightly constricted at the ends; striae 7-9 in 10 µ, coarse, more or less irregularly disposed.

## Genus: *Achnanthes* Bory

### 14) *Achnanthes inflata* (Kuetz.) Grun. (Pl. 30, Fig. 1)

Sarode and Kamat, 1984, p 55, pl. 5, f. 113

Frustules broadly linear, bent in the middle in girdle view; valves 42.8-55.8 µ long; 12.2 -16µ broad, linear, dilated in the middle, ends broadly rounded; raphe valve with thin and straight raphe; axial area narrow; central area stauroid; striae 11-12 in 10 µ, radial and punctuate, rapheless valves with narrow, unilaterally disposed pseudoraphe, without central area; striae 9-10 in 10 µ, punctate.

### 15) *A. inflata* (Kuetz.) Grun. f. *vidarbhensis* Sarode et Kamat (Pl. 30, Fig. 2)

Sarode and Kamat, 1984, p 55, pl. 5, f 115

Valves 89-95.5μ long; 16-16.6μ broad, linear trigibbous with broadly cuneate rounded ends; raphe valve with thin and straight raphe, axial area fairly broad, linear, central area stauroid, reaching the margins; striae 9-10 in 10 μ, radial and punctuate; rapheless valve with unilateral, narrow pseudoraphe; central area absent; striae 9 in 10 μ, coarsely punctate, curved at the ends.

## Sub order: Naviculineae

## Family: Naviculaceae

## Genus: *Navicula* Bory

### 16) *Navicula cari* Ehr. (Pl. 30, Fig. 3)

Sarode and Kamat, 1984, p 104, pl. 11, f. 245

Valves 17.5-20 μ long, 4.2-4.5 μ broad, narrowly lanceolate with almost acute rounded ends; raphe thin and straight with curved terminal fissures; axial area narrow, linear, central area large, striae 18 in 10 μ, radial in the middle and convergent at the ends.

### 17) *N. cari* Ehr. v. *angusta* Grun. (Pl. 30, Fig. 4)

Sarode and Kamat, 1984, p 104, pl. 11, f. 247

Valves 35 μ long, 7.5 μ broad, narrowly lanceolate with slightly constricted or unconstricted, broadly rounded ends; raphe thin and straight with curved terminal fissures; axial area narrow, linear; central area large, quadrate; striae 10-12 in 10 μ, indistinctly lineate, radial in the middle and convergent at the ends, middle striae shorter.

### 18) *N. cari* Ehr. f. *indica* f. Sarode and Kamat (Pl. 30, Fig. 5)

Sarode and Kamat, 1984, p 104, pl. 11, f. 246

Valves 40 μ long, 7.5 μ broad, narrowly lanceolate, tapering towards the ends with somewhat produced and broadly rounded ends; raphe thin and straight; axial area narrow; central area large, transversely extended, quadrangular; striae 12 in 10 μ, strongly radial in the middle and convergent towards the ends, middle striae shorter.

### 19) *N. cryptocephala* Kuetz. (Pl. 30, Fig. 6)

Sarode and Kamat, 1984, p 106, pl. 12, f. 254

Valves 22-35.5 μ long 8.5-9.2 μ broad, lanceolate with produced, rounded ends; raphe thin and straight; axial area narrow; central area slightly extended transversely, small; striae 10-12 in 10 μ, radial in the middle and slightly convergent at the ends.

### 20) *N. cuspidata* Kuetz. (Pl. 30, Fig. 7)

Sarode and Kamat, 1984, p 107, pl. 12, f. 258

Valves 65-110 μ long, 14.6-20 μ broad, rhombic lanceolate with acutely rounded ends; raphe thin and straight with hooked unilaterally bent central pores and large terminal fissures; axial area narrow, linear, central area very small, striae transverse, 14-16 in 10 μ, parallel, slightly convergent at the ends. Longitudinal striae about 25 in 10 μ.

### 21) *N. cuspidata* Kuetz. v. *ambigua* (Ehr.) Cleve f. *diminuta.* A.Cl. (Pl. 30, Fig. 8)

Sarode and Kamat, 1984, p 108, pl. 12, f. 261

Valves 60-85μ long, 16.1-20.5μ broad, broadly lanceolate with constricted, rounded ends; raphe thin and straight with unilaterally bent central pores; axial area narrow, transverse striae 14-16 in 10 μ, longitudinal striae 20-22 in 10μ, fine.

### 22) *N. cuspidata* Kuetz. f. *brevirostrata* Gandhi (Pl. 30, Fig. 9)

Sarode and Kamat, 1984, p 107, pl. 12, f. 259

Valves 55.8-65.5 μ long, 17-27 μ broad, elliptic lanceolate with constricted shortly rostrate, subtruncate ends; raphe thin and straight with central pores hook like; axial area very narrow, linear central area very small, transverse striae 14-16 in 10 μ, longitudinal striae 24-26 in 10 μ, fine.

### 23) *N. cuspidata* Kuetz. v. *heribaudi* Peragallo (Pl. 30, Fig. 10)

Sarode and Kamat, 1984, p 109, pl. 12, f. 263

Valves 106.4-173.9 μ long, 24.9-35.4 μ broad, rhombic lanceolate with narrowly attenuated ends; raphe thin and straight with unilaterally bent central pores; axial area narrow, central area slightly widened, transverse striae 18-20 in 10 μ, strongly radial, longitudinal striae more numerous and faint, indistinct.

### 24) *N. dicephala* (Ehr.) W. Smith v. *sphaerophora* A. Cl. (Pl. 30, Fig. 11)

Sarode and Kamat, 1984, p 110, pl. 13, f. 268

Valves 20-25 μ long, 7-8 μ broad, linear or linear elliptical with constricted, capitate, rounded ends; raphe thin and straight; axial area narrow, linear; central area fairly large, somewhat roundish, striae 14-15 in 10 μ, radial throughout and curved.

### 25) N. *feverborni* (Feverb.) Hustedt (Pl. 30, Fig. 12)

Sarode and Kamat, 1984, p 111, pl. 13, f 271

Valves 35.5-41 μ long; 6.5-7.5 μ broad, linear lanceolate with broadly rounded ends; raphe thin and straight, axial area very narrow; central area small, unilateral, striae 14 in 10 μ, strong and strongly radial.

### 26) N. *iniqua* Krasske (Pl. 30, fig.13)

Sarode and Kamat, 1984, p 113, pl. 13,f. 280

Valves 16-18 μ long, 4.5-7.6 μ broad, linear with somewhat sinuous,wedge shaped and acutely rounded ends, raphe thin and straight; axial area broad central area rounded, striae 18-20 in 10 μ, radial.

### 27) N. *laterostrata* Hustedt (Pl. 30, Fig. 14)

Sarode and Kamat, 1984, p 113, pl. 13, f. 281

Valves 18.5-28.5 μ long, 6.6-9.5 μ broad, elliptic lanceolate with broadly rounded and more or less capitate ends; raphe thin and straight; axial area very narrow; central area large, rhombic to rounded; striae 14-16 in 10 μ in the middle and up to 20 in 10 μ at the ends, delicate and radial.

### 28) N. *mutica* Kuetz (Pl. 31, Fig. 1)

Sarode and Kamat, 1984, p 115, pl. 13, f. 286

Valves 13.5-20.8 μ long, 5.6-6.5 broad, elliptic lanceolate with very slightly constricted, broadly rounded ends; raphe thin and straight, axial area narrow; central area large, striae 20-24 in 10 μ, radial, clearly punctuate and somewhat closer at the ends.

### 29) N. *pupula* Kuetz v. *elliptica* Hustedt. (Pl. 31, Fig. 2)

Sarode and Kamat, 1984, p 118, pl. 13, f. 297

Valves 12.5-16 μ long, 6.2-6.8 μ broad, broadly elliptic lanceolate with rounded ends; raphe thin and straight, axial area narrow; linear, central area rectangular, large, striae 22-24 in 10 μ, fine, radial, curved, long and short striae alternating in the middle.

### 30) N. *pupula* Kuetz v. *rostrata* Hustedt. (Pl. 31, Fig. 3)

Sarode and Kamat, 1984, p 119, pl. 13, f. 299

Valves 15-32 μ long, 6.6-8.5 μ broad, elliptic with narrowed constricted rostrate ends; raphe thin and straight; polar areas distinct; axial area very narrow; central area large rectangular, striae 20-22 in 10 μ, radial and fine.

### 31) *N. radiosa* Kuetz. v. *tenella* (breb. Ex kuetz.) Grun. (Pl. 31, Fig. 4)

Sarode and Kamat, 1984, p 120, pl. 14, f. 306

Valves 27.5-34 µ long, 5.6-6.5 µ broad, narrowly lanceolate and gradually tapering to somewhat acutely rounded ends, raphe thin and straight; axial area narrow, central area large, rounded, striae 14-16 in 10 µ, radial in the middle.

### 32) *N. reinhardtii* Grun. f. *gracilior* Grun. (Pl. 31, Fig. 5)

Sarode and Kamat, 1984, p 121, pl. 14, f. 307

Valves 50 µ long, 10 µ broad, elliptic, lanceolate with, broadly rounded ends; raphe somewhat thickened and straight with distinct central pores; axial area narrow; central area large, rectangular; striae 9-10 in 10 µ, radial throughout.

### 33) *N. rhynchocephala* Kuetz. v. *elongata* Mayer. (Pl. 31, Fig. 6)

Sarode and Kamat, 1984, p 121, pl. 14, f. 310

Valves 45.1-55 µ long, 7-8 µ broad, oblong, slender, with slightly dilated capitate ends; raphe thin and straight, axial area narrow, linear, central area rounded, striae 14-16 in 10 µ, denser towards the ends.

### 34) *N. schonfeldii* Hustedt. (Pl. 31, Fig. 7)

Sarode and Kamat, 1984, p 123, pl. 14, f. 316

Valves 15.8-21 µ long, 6.8-7.2 µ broad, elliptical or elliptical lanceolate with rounded ends; raphe thin and straight; axial area very narrow; central area rectangular; striae 14-15 in 10 µ, radial, short and long striae alternating in the middle.

### 35) *N. varidula* Kuetz. v. *capitata* Mayer (Pl. 31, Fig. 8)

Sarode and Kamat, 1984, p.126, pl. 14, f.328

Valves 28.8-41 µ long, 7.5-8.5 µ broad, linear elliptical with somewhat produced, capitate rounded ends; raphe thin and straight; axial area narrow, linear; central area moderately wide, quadrate; striae 10-12 in 10 µ, distinctly lineate, radial in the middle and convergent at the ends.

## Genus : *Anomoeoneis* E. Pfitzer

### 36) *Anomoeoneis lanceolata* Gandhi (Pl. 31, Fig. 9)

Sarode and Kamat, 1984, p 97, pl. 11, f. 235

Valves 48-64.8 µ long, 14-16.2 µ broad, somewhat broadly lanceolate, robust with rostrate ends; raphe thin and straight with prominent central pores and curved terminal fissures, axial area narrow, linear; central area large, unilateral;

striae 18-20 in 10 µ, slightly radial, coarsely punctate, crossed by many hyaline longitudinal irregular spaces.

### 37) *A. sculpta* (Ehr.) Cleve (Pl. 31, Fig. 10)

Sarode and Kamat, 1984, p 98, pl. 11, f. 236

Valves 28.5-57.6 µ long, 11.5-14.4 µ broad, elliptic lanceolate, robust with constricted rostrate ends; raphe thin and straight with central pores unilaterally bent and terminal fissures slightly curved, axial area narrow, linear; central area large, unilaterally expanded with very fine indistinct punctae, irregualry disposed; striae 18-20 in 10 µ, radial, coarsely punctate, crossed by many hyaline longitudinal irregular spaces.

## Genus: *Stauroneis* Ehr.

### 38) *Stauroneis anceps* Ehr. v. *hyalina* Brun. et Perag. (Pl. 31, Fig. 11)

Sarode and Kamat, 1984, p 90, pl. 10, f. 213

Valves 55 µ long, 10 µ broad, subelliptical lanceolate with produced, feebly capitate ends; raphe thin and straight, central pores distinct and terminal fissures very slightly curved; axial area very narrow; central area large, narrowly rectangular and reaching the margins; striae 30 in 10 µ, radial and very finely punctate.

### 39) *S. kirtikarii* Sarode and Kamat (Pl. 31, Fig. 12)

Sarode and Kamat, 1984, p 91, pl. 11, f. 230

Valves 32.5-38.4 µ long, 8.1-10 µ broad, elliptic to linear elliptic with broadly rostrate, very slightly capitate ends; raphe thin and straight with distinct central pores; axial area narrow, central area large, stauroid, narrowed towards the margins; striae 22-24 in 10 µ, very fine and finely punctate.

### 40) *S. phoenicenteron* Ehr. f. *producta* Gandhi (Pl. 31, Fig. 13)

Sarode and Kamat, 1984, p 94, pl. 11, f. 226

Valves 40.3-91.5 µ long, 8.6-18 broad, narrowly lanceolate, with slightly constricted, produced rounded ends; raphe thick with distinct central pores and curved terminal fissures; axial area narrow; central area a linear stauros; striae 18-24 in 10 µ, radial, fine and distinctly punctate.

## Genus: *Neidium* Pfitzer

### 41) *Neidium affine* (Ehr.) Cleve v. *amphirhynchus* (Ehr.) Cleve f. *truncatula* Gonzalves et Gandhi (Pl. 31, Fig. 14)

Sarode and Kamat, 1984, p 78, pl. 9, f. 179

Cells usually solitary and free floating, rectangular in girdle view, without intercalary bands, valves 39-50 μ long, 10-12 μ broad, linear with almost parallel margins, suddenly contracted towards the ends and produced into broadly truncate poles, raphe thin and straight, central pores bent in opposite directions and terminal fissures narrowly bifurcated; axial area narrow; central area small and rhomboid; striae 20-22 in 10 μ, fine but clearly punctate, crossed by a hyaline furrow near the margins.

### 42) *Neidium amphigomphous* (Ehr.) Cleve v. *obtusum* A.Cl. (Pl. 31, Fig. 15)

Sarode and Kamat, 1984, p 79, pl. 9, f. 181

Valves 37.5 μ long, 10 μ broad, elliptic lanceolate with obtuse, rounded ends; raphe thin and straight with central pores bent in opposite directions; axial area narrow; central area large, slightly transversely widened; striae 18-20 in 10 μ, finely punctate, crossed by more than two longitudinal furrows.

### 43) *N. iridis* (Ehr.) Cleve f. *ambigua* Gonzalves et. Gandhi (Pl. 31, Fig. 16)

Sarode and Kamat, 1984, p 83, pl. 9, f. 193

Valves 82.5 μ long, 17.5 μ broad, linear with almost parallel margins and broadly rounded, distinctly wedge shaped ends; raphe thin and straight with central pores bent in opposite directions; axial area narrow; central area transversely elliptical; striae 16-18 in 10 μ.

### 44) *N. iridis* (Ehr.) Cleve f. *dhulensis* Sarode and Kamat (Pl. 31, Fig. 17)

Sarode and Kamat, 1984, p 83, pl. 9, f. 194

Valves 40 μ long, 10 μ broad, linear elliptical to linear lanceolate with almost parallel margins and broadly rounded, somewhat wedge shaped ends; raphe thin and straight with central pores bent in opposite directions and terminal fissures bifurcated; striae 28-30 in 10 μ, finely punctate, radial, crossed by a marginal furrow.

## Genus: *Diploneis* Ehr.

### 45) *Diploneis elliptica* (Kuetz.) Cleve v. *ladogensis* Cleve (Pl. 32, Fig. 1)

Sarode and Kamat, 1984, p 86, pl. 10, f. 202

Valves 52.5 μ long, 30 μ broad, elliptical to rhombic elliptical; raphe between the ribs; ribs widened in the central nodule; axial area very narrow; central area slightly dilated, elliptical; costae 9 in 10 μ, radial at the ends, alternating with a single row of very coarse punctate or alveoli; alveoli 8-9 in 10 μ.

### 46) *D. puella* (Schum.) Cleve (Pl. 32, Fig. 2)

Sarode and Kamat, 1984, p 87, pl. 10, f 205

Valves 15 μ long, 7.5 μ broad, elliptic with rounded ends, valve surface costate; costae 13-14 in 10 μ, thick and distinct, continued into the furrows as large alveoli; central nodule large, dilated.

## Genus: *Caloneis* Cleve

### 47) *Caloneis beccariana* Grun. (Pl. 32, Fig. 3)

Sarode and Kamat, 1984, p 72, pl. 8, f. 162

Valves 30-51 μ long, 7.1-9 μ broad, lanceolate to linear lanceolate, somewhat inflated in the middle, broadly rounded ends; axial area fairly wide; central area quite large and reaching the margins, sometimes more widened on one side than the other; striae 21-23 in 10 μ, conspicusly radial but clearly convergent at the ends, crossed by a fine longitudinal line near the margin.

### 48) *C. beccariana* Grun. v. *hustedtii* Gandhi (Pl. 32, Fig. 4)

Sarode and Kamat, 1984, p 72, pl. 8, f 163

Valves 26.5-30 μ long; 7.1-7.8 μ broad, lanceolate to linear lanceolate, slightly inflated in the middle, with broadly rounded ends; axial area moderate; central area quite large and reaching the margins; striae 22-24 in 10 μ, slightly radial, crossed by a fine longitudinal line near the margin.

### 49) *C. permagna* (Bail.) Cleve (Pl. 32, Fig. 5)

Sarode and Kamat, 1984, p 72, pl. 8, f 164

Frustules large and robust; valves 70-132 μ long, 27.4-41 μ broad, rhombic lanceolate with somewhat produced, broadly rounded ends; raphe thick and straight; central pores large, unilaterally bent, terminal fissures broadly curved and clear; axial area large, moderately lanceolate; central area large, circular and somewhat unilateral; striae 13-14 in 10 μ, radial in the middle and convergent at the ends; crossed by two longitudinal lines slightly away from the margin.

### 50) *C. silicula* (Ehr.) Cleve (Pl. 32, Fig. 6)

Sarode and Kamat, 1984, p 74, pl. 8, f. 168

Valves 35-53.5μ long, 6.5-8μ broad, linear to lanceolate with triundulate margins and broadly wedge shaped ends, raphe thin and straight, axial area moderate, linear lanceolate; central area large; striae 20-22 in 10μ, slightly radial throughout, crossed by a fine longitudinal line near the margin.

### 51) *C. silicula* (Ehr.) Cleve v. *intermedia* Mayer (Pl. 32, Fig. 7)

Sarode and Kamat, 1984, p 75, pl. 8, f. 172

Valves 56.1-65µ long, 11.5-12µ broad; linear, feebly undulate, slightly inflated in the middle and near the ends; ends somewhat cuneate rounded; raphe thin and straight with central pores distinct and curved terminal fissures; axial area broadly lanceolate; central area reaching the sides; striae 24-26 in 10µ, slightly radial in the middle and perpendicular to the middle line towards the ends.

### 52) *C. silicula* (Ehr.) Cleve v. *tenuis* (Hustedt.) Mayer (Pl. 32, Fig. 8)

Sarode and Kamat, 1984, p 76, pl. 8, f. 175

Valves 50.5-51µ long, 9-9.5µ broad, linear, very slightly gibbous in the middle with broadly rounded ends; raphe thin and straight; axial area moderate; central area large, reaching the margins; striae 26 in 10µ, fine radial, crossed by a fine longitudinal line near the margin.

### 53) *C. ventricosa* (Ehr.) Meist. v. *alpina* (Cleve) Patr. (Pl. 32, Fig. 9)

Sarode and Kamat, 1984, p 77, pl. 8, f. 178

Valves 50 µ long, 12.5 µ broad, broadly linear, gibbous in the middle with rounded poles; raphe thin and straight, terminal fissures hooked in the same directions; axial area narrow, widening towards the middle; central area elliptical; striae 16-18 in 10 µ, parallel in the middle and becoming radial near the poles, crossed by a longitudinal line away from the margins.

## Genus: *Pinnularia* Ehr.

### 54) *Pinnularia acrosphaeria* (Breb.) W. Smith (Pl. 33, Fig. 1)

Sarode and Kamat 1984, p 133, pl. 15, f. 340.

Valves 36.5 µ long, 6.5 µ broad, linear, slightly inflated in the middle and at the ends; raphe thin and straight with closely set, axial area very wide with irregularly disposed punctae; striae 12-14 in 10 µ, very slightly radial.

### 55) *P. acrosphaeria* (Breb.) W. Smith v. *minor* Cleve (Pl. 33, Fig. 2)

Sarode and Kamat, 1984, p 134, pl. 15, f. 342

Valves 34.5-44 µ long, 7.5-9.5 µ broad, very slightly swollen in the middle with broadly rounded ends; raphe thin and straight with unilaterally bent central pores; axial area wide; striae 12-14 in 10 µ slightly radial in the middle.

### 56) *P. acrosphaeria* (Breb.) W. Smith f. *undulata* Cleve (Pl. 33, Fig. 3)

Sarode and Kamat, 1984, p 133-134, pl. 15, f. 341

Valves 56.5-72 µ long, 10-11.5 µ broad, linear, inflated in the middle with broadly rounded ends; raphe thin and straight with closely set, axial area very wide, with irregularly disposed punctae; central area not prominent; striae 12-

14 in 10 μ slightly radial in the middle.

## 57) *P. aestuarii* Cleve v. *interrupta* (Hustedt) A.Cl. (Pl. 33, Fig. 4)

Sarode and Kamat, 1984, p 134, pl. 15, f. 344

Valves 62-82 μ long, 9.5-11 μ broad, linear, with broadly subcuneate ends; raphe thick, complex, with large central pores, unilaterally bent and semicircular terminal fissures; axial area ¼ the breadth of the valve, linear; central area fairly large, reaching the margins; striae 9-10 in 10 μ, radial in the middle and convergent at the ends.

## 58) *P. borealis* Ehr. v. *lonavlensis* Gandhi (Pl. 33, Fig. 5)

Sarode and Kamat, 1984, p 136, pl. 15, f. 349

Valves 65-84 μ long, 16.5-18 μ broad, linear to linear lanceolate, with constriction in the middle; ends tapering and rounded; raphe thick, complex with unilaterally curved central pores and bayonet shaped terminal fissures; axial area wide, linear; central area rounded; striae 4-6 in 10 μ, strong and radial, slightly convergent at the ends.

## 59) *P. brevicostata* Cleve v. *indica* Gandhi (Pl. 33, Fig. 6)

Sarode and Kamat, 1984, p 138, pl 15, f. 357

Valves 87.5-97 μ long, 19.5-21 μ broad, linear elliptical with rounded ends; raphe thick with central pores unilaterally bent, axial area wide, lanceolate; striae 6 in 10 μ, very coarse.

## 60) *P. cardinaliculus* (Cleve) Lund. (Pl. 33, Fig. 7)

Sarode and Kamat, 1984, p 138, pl. 16, f. 358

Valves 64-79.2 μ long, 10.3-16.2 μ broad, linear with very weakly inflated middle part and broadly rounded ends; raphe slightly thickened and straight with unilaterally bent central pores and curved terminal fissures; axial area fairly wide, central area quite large, reaching the margins; striae 8-9 in 10 μ, thick, with faint narrow, longitudinal band, radial in the middle and convergent at the ends.

## 61) *P. dolosa* Gandhi (Pl. 33, Fig. 8)

Sarode and Kamat, 1984, p 141, pl. 16, f. 367

Valves 85 μ long, 15 μ broad, linear, tumid in middle with slightly swollen broadly rounded ends; raphe formed in the hyaline zone, central pores closely set and unilaterally bent, terminal fissures curved; axial area wide, with fine irregularly disposed punctae; central area slightly unilaterally dilated; striae 9-11 in 10 μ, thick slightly radial in the middle and convergent at the ends.

### 62) *P. kiusiuensis* Skv. (Pl. 33, Fig. 9)

Sarode and Kamat, 1984, p 145, pl. 17, f. 380

Valves 75 μ long, 12.5 μ broad, linear, slightly gibbous in the middle with swollen rounded ends; raphe thick, more or less complex with central pores unilaterally bent and terminal fissures thick, curved; axial area linear; central area large and unilaterally reaching the margins; striae 9-10 in 10 μ, thick slightly radial in the middle and convergent at the ends, with longitudinal band.

### 63) *P. kolhapurensis* Gandhi (Pl. 33, Fig. 10)

Sarode and Kamat, 1984, p 145, pl. 17, f. 381

Valves 32.5-37 μ long, 7.5-8 μ broad, sublinear with feeble convex margins, truncate rounded ends; raphe thin and straight with unilaterally bent central pores and slightly curved terminal fissures; axial area linear; central area wide, rhomboid, striae thick, 12-13 in 10 μ, radial in the middle and convergent at the ends.

### 64) *P. legumen* Ehr. v. *florentina* (Grun.) Cleve (Pl. 33, Fig. 11)

Sarode and Kamat, 1984, p 145-146, pl. 17, f. 383

Valves 64.2-75 μ long, 12.7-13 μ broad, linear lanceolate with triundulate margins and constricted, slightly produced, capitate ends; raphe thick somewhat undulated; axial area wide, linear; central area large, striae 8-11 in 10 μ, thick strongly radial in the middle and convergent at the ends.

### 65) *P. major* (Kuetz.) Cleve v. *linearis* Cleve (Pl. 33, Fig. 12)

Sarode and Kamat, 1984, p 147, pl. 17, f. 387

Valves 95-126 μ long, 16-25.2μ broad, linear with very slightly inflated margins in the middle part and somewhat cuneately rounded ends; raphe thick and complex with unilaterally bent central pores and bayonet shaped, curved terminal fissures; axial area fairly wide; central area broader; striae 7-8 in 10 μ, slightly radial in the middle, convergent at the ends.

### 66) *P. marathwadensis* Sarode and Kamat (Pl. 34, Fig. 1)

Sarode and Kamat, 1984, p 148, pl. 19, f. 426

Valves 95-104 μ long, 12.6-13 μ broad, linear, slightly gibbous in the middle with broadly rounded ends; raphe thick, with thick unilaterally bent central pores and slightly curved semicircular terminal fissures; axial area wide, about 1/5-1/4 the breadth of the valve; central area large, rhomboid, narrowly reaching the margins; striae 10-14 in 10 μ, radial and slightly curved in the middle and convergent at the ends.

## 67) *P. meisteri* A.Cl. v. *sandica* A.Cl. (Pl. 34, Fig. 2)

Sarode and Kamat, 1984, p 148, pl. 17, f. 389

Valves 60 µ long, 12.5 µ broad, broadly linear, slightly swollen in the middle with somewhat inflated rounded ends; raphe thin and straight with terminal fissures curved; axial area narrow, linear; central area only slightly dilated; striae about 12 in 10 µ, coarse, slightly radial in the middle and slightly convergent at the end

## 68) *P. microstauron* (Ehr.) Cleve (Pl. 34, Fig. 3)

Sarode and Kamat, 1984, p 149, pl. 17, f. 392

Valves 47-55.5 µ long, 8.5-10 µ broad, linear lanceolate with constricted, somewhat captitate ends; raphe thin and slightly undulate; terminal fissures slightly curved; axial area moderate, lanceolate, central area large, reaching the margins, striae 12-14 in 10 µ, strong, radial in the middle and convergent at the ends.

## 69) *P. notata* (Perag. et. Her.) A. Cl. v. *rostrata* A. Cl. (Pl. 34, Fig. 4)

Sarode and Kamat, 1984, p 151, pl. 18, f. 399

Valves 18-33 µ long, 4.5-7.2 µ broad, linear with somewhat abruptly constricted, produced rounded ends; raphe thin and straight; axial area narrow, linear; central area large, striae 12-13 in 10 µ, radial in the middle and convergent at the ends.

## 70) *P. pseudoluculenta* Gandhi (Pl. 34, Fig. 5)

Sarode and Kamat, 1984, p 152, pl. 18, f. 404

Valves 126-137 µ long, 17.5-18 µ broad, narrowly linear with concave margins in between the middle and the ends with gradually swollen broadly rounded ends; raphe thick, complex with central pores unilaterally bent and terminal fissures obliquely curved; axial area about 1/3 the breadth of the valve, linear lanceolate; central area quadrate; striae 7-8 in 10 µ, thick, slightly radial in the middle and convergent at the ends with distinct narrow bands.

## 71) *P. stauroptera* (Rabh.) Cleve v. *parva* Grun. (Pl. 34, Fig. 6)

Sarode and Kamat, 1984, p 154, pl. 18, f. 412

Valves 63.5-72 µ long, 10.3-11.9 µ broad, with slightly convex margins and very slightly constricted, somewhat wedge shaped capitate ends; raphe thin and straight with unilaterally bent central pores and curved terminal fissures, axial area wide; central area large, rhomboid, reaching the margins; striae 9-10 in 10 µ, radial in the middle and convergent at the end.

### 72) *P. stomatophoroides* Mayer v. *ornata* A.Cl. f. *erlangensis* Mayer (Pl. 34, Fig. 7)

Sarode and Kamat, 1984, p 155, pl. 18, f. 414

Valves 67.5-95.7 μ long, 7.5-12.5 μ broad, linear or sublinear with feebly triundulate margins and broadly rounded, raphe thick, central area large, reaching the margins with arcuate row of coarse punctae on either side of the central pores, striae 11-12 in 10 μ, thick strongly radial in the middle and convergent at the ends.

### 73) *P. sudetica* Hilse v. *commutata* (Grun.) Cleve f. *obtusata* Gandhi (Pl. 34, Fig. 8)

Sarode and Kamat, 1984, p 157, pl. 19, f. 419

Valves 50-60 μ long, 9-9.5 μ broad, linear with almost parallel margins and broadly rounded ends; raphe thick with unilaterally curved pores and curved terminal fissures; axial area narrow, linear central area slightly widened, striae 11-13 in 10 μ, slightly radial in the middle.

## Genus: *Mastogloia* Thwaites

### 74) *Mastogloia baltica* Grun. (Pl. 34, Fig. 9)

Sarode and Kamat, 1984, p 60, pl. 6, f. 128.

Valves 34-47.5 μ long; 12-13.5 μ broad, elliptic lanceolate with produced capitate ends; raphe nearly straight in between two longitudinal ribs; axial area narrow; central area small; roundish; striae 20-21 in 10 μ slightly radial and finely punctate.

### 75) *M. recta* Hustedt. (Pl. 34, Fig. 10)

Sarode and Kamat, 1984, p 61, pl. 6, f. 131

Valves 45 μ long, 12.5 μ broad, linear elliptical to linear lanceolate with slightly constricted, produced, obtusely rounded ends; raphe thick, complex with slightly unilaterally bent central pores; axial area narrow, linear; central area fairly large, rounded; striae 12-14 in 10 μ, radial throughout or somewhat convergent at the extreme ends.

## Genus: *Frustulia* Ag.

### 76) *Frustulia jogensis* Gandhi (Pl. 34, Fig. 11)

Sarode and Kamat, 1984, p 63, pl. 6, f. 138

Valves 57.5 μ long, 12.5 μ broad, linear lanceolate with constricted, produced beak like rounded ends; raphe thin, delicate and straight, enclosed between the siliceous ribs; axial area very narrow, linear; central area small; striae 32 in 10 μ, perpendicular to the middle line.

## Genus: *Gyrosigma* Hassall

### 77) *Gyrosigma acuminatum* (Kuetz.) Rabh. (Pl. 34, Fig. 12)

Sarode and Kamat, 1984, p 66, pl. 7, f. 145

Frustules solitary; valves 99-106.5 µ long, 10-12.8 µ broad, sigmoid, lanceolate in outline, gradually tapering from the middle towards the ends; ends broadly rounded; raphe sigmoid and central; axial area narrow; central area small and elliptical; striae 18 in 10 µ; transverse and longitudinal striae at equal distances from one another.

### 78) *G. bhusavalensis* Sarode and Kamat (Pl. 34, Fig. 13)

Sarode and Kamat, 1984, p 67, pl. 7, f. 153

Valves 115-127.8 µ long, 16.2-23.5 µ broad, slightly sigmoid, linear, attenuated towards the poles with obliquely rounded ends; raphe central and sigmoid; axial area very narrow; central area small elliptical; transverse striae 14-15 in 10 µ, coarse, longitudinal striae 17-18 in 10 µ, distinct.

### 79) *G. khandeshensis* Sarode and Kamat (Pl. 34, Fig. 14)

Sarode and Kamat, 1984, p 67, pl. 17, f. 154

Frustules free floating, solitary; valves 91-100.8 µ long, 13.6-18 µ broad, slightly sigmoid, linear lanceolate with broadly rounded ends; raphe central, slightly sigmoid, axial area narrow; central area small and elliptical; transverse striae 22-24 in 10 µ, slightly radial, fine but distinct; longitudinal striae 18-20 in 10 µ, fine.

### 80) *G. kuetzingii* (Grun.) Cleve (Pl. 34, Fig. 15)

Sarode and Kamat, 1984, p 68, pl. 7, f. 149

Valves 85.4-93 µ long, 12.8-14.4 µ broad, slightly sigmoid, lanceolate with rounded ends; raphe in the centre, slightly sigmoid; axial area narrow, linear, central area small; transverse striae 24 in 10 µ, coarse; longitudinal striae 26 in 10µ, fine.

## Family: Gomphonemaceae

## Genus: *Gomphonema* Ag.

### 81) *Gomphonema aequatoriale* Hustedt. (Pl. 35, Fig. 1)

Sarode and Kamat, 1984, p 182, pl. 21, f.486

Valves 55 µ long, 10 µ broad, clavate in the middle inflated with broadly rounded apex and narrowly produced rounded base; raphe thick; axial area narrow; central area widened; striae 9-10 in middle and 11 in 10 µ, towards the ends, clearly punctate.

## 82) *G. balatonis* Pant. (Pl. 35, Fig. 2)

Sarode and Kamat, 1984, p 183, pl. 21, f. 489

Valves 90-92 μ long, 11.5-12 μ broad, narrowly lanceolate clavate with rounded ends, without constrictions; raphe thick and straight, terminal fissures distinct; axial area narrow, linear, central area slightly unilaterally enlarged with an isolated, stigma on the opposite side; striae 6-8 in 10 μ, radial, fine and clearly punctate.

## 83) *G. gracile* Ehr. v. *intricatiforme* Mayer (Pl. 35, Fig. 3)

Sarode and Kamat 1984, p 186, pl. 22, f. 500

Valves 56.5-66.5μ long, 10.5-11.3μ broad, lanceolate, clavate inflated in the middle; ends obtuse, rounded; raphe somewhat thick and straight with distinct central nodules and curved terminal fissures; axial area narrow, linear; central area slightly expanded with an isolated stigma on the opposite side; striae 10-12 in 10 μ in the middle and up to 14 in 10 μ, towards the ends, slightly radial and coarse.

## 84) *G. gracile* Ehr. v. *major* Grun. (Pl. 35, Fig. 4)

Sarode and Kamat, 1984, p 187, pl. 22, f. 502

Valves 67.5 μ long, 10 μ broad, narrowly lanceolate, turgid in the middle and gradually tapering towards the rounded apex and base; raphe thin and straight; axial area narrow; central area with an isolated stigma on the opposite side; striae 10-12 in 10 μ slightly radial and finely punctate.

## 85) *G. intricatum* Kuetz. (Pl. 35, Fig. 5)

Sarode and Kamat, 1984, p 188, pl. 22, f. 506

Valves 33-45 μ long, 7-7.9 μ broad, subclavate with constricted, slightly swollen broadly rounded apex and attenuated rounded base; raphe slightly thick; axial area narrow, linear; central area unilateral with an isolated stigma on the opposite side; striae 8 in 10 μ, in the middle and 10 in 10 μ, towards the ends, radial and coarse.

## 86) *G. laneolatum* Ehr. (Pl. 35, Fig. 6)

Sarode and Kamat, 1984, p 190, pl. 22, f. 514

Valves 40-110 μ long, 8-16 μ broad, lanceolate clavate with distinctly rounded apex and base, base somewhat narrower; raphe slightly thick and straight; axial area narrow, linear, central area slightly unilateral with an isolated stigma on the opposite side; striae 11-12 in 10 μ, radial and lineate.

## 87) *G. lanceolatum* Ehr. v. *insignis* (Greg.) Cleve (Pl. 35, Fig. 7)

Sarode and Kamat, 1984, p 191, p.l 22, f. 516

Valves 29-45 µ long, 7-9 µ broad, broadly lanceolate clavate with acute ends; raphe slightly thick with unilaterally bent, central pores and distinct terminal fissures; axial area moderate, linear; central area unilateral with an isolated stigma on the opposite side; striae 10-12 in 10 µ, radial, fine and distinctly punctate.

### 88) *G. longiceps* Ehr. v. *subclavata* Grun. (Pl. 35, Fig. 8)

Sarode and Kamat, 1984, p 191, pl. 22, f. 518

Valves 28.4-44.6 µ long, 8.5-10.8 µ broad, lanceolate, slightly tumid in the middle with somewhat broadly rounded apex and attenuated base; raphe thin and straight; axial area narrow, central area with an isolated stigma on the opposite side; striae 9-10 in 10 µ, almost parallel in the middle and slightly radial towards the ends, coarsely punctate.

### 89) *G. magnifica* Gandhi (Pl. 35, Fig. 9)

Sarode and Kamat, 1984, p 192, pl. 23, f. 519

Valves 72-82.8 µ long, 17-19 µ broad, lanceolate, clavate, somewhat rhomboid, with acuately rounded apices; raphe thick and straight, axial area narrow, linear; central area small, slightly unilateral with an isolated stigma on the opposite side; striae 6-8 in 10 µ, radial, distinctly punctate, distantly placed in the middle.

### 90) *G. moniliforme* Gandhi (Pl. 35, Fig. 10)

Sarode and Kamat, 1984, p 193, pl. 23, f. 522

Valves 52-62 µ long, 18.5-20.5 µ broad, broadly clavate, apex constricted; produced, sub-truncate; base with slightly concave margins and attenuated; raphe thick, central pores unilaterally bent and terminal fissures distinct; axial area narrow, linear; central area small, unilaterally dilated with an isolated stigma on the opposite side; striae 7-8 in 10 µ, radial, coarsely punctate.

### 91) *G. subventricosum* Hustedt. (Pl. 35, Fig. 11)

Sarode and Kamat, 1984, p 199, pl. 23, f. 542

Valves 45-70 µ long, 12.3-15.5 µ broad, lanceolate clavate with somewhat produced, slightly constricted; feebly capitate apex and attenuated base; raphe thin and straight; axial area narrow; central area unilateral with an isolated stigma on the opposite side; striae 10-12 in 10 µ, radial.

## Family: Cymbellaceae

## Genus: *Amphora* Ehr.

### 92) *Amphora costata* W. Smith (Pl. 35, Fig. 12)

Sarode and Kamat, 1984, p 160, pl. 19, f. 431

Frustules linear elliptical in girdle view; valves 25 µ long 5 µ broad, strongly arcuate on the dorsal margin and slightly concave more or less straight on the ventral margin; ends strongly constricted, capitate; raphe thin, more or less straight, slightly directed to the ventral side towards the ends; striae about 15 in 10 µ, coarsely punctate.

### 93) *A. maharashtrensis* Sarode and Kamat (Pl. 35, Fig. 13)

Sarode and Kamat, 1984, p 161, pl. 19, f. 438

Frustules elliptic; valves 12.5 µ long, 5 µ broad, strongly convex on the dorsal side and more or less straight on the ventral side with inwardly bent rounded ends; raphe thin close to ventral margin, more or less straight, terminal fissures very slightly directed towards the ventral margin; striae 10-22 in 26 µ, coarsely punctate in the middle.

### 94) *A. ovalis* Kuetz. v. *gracilis* (Ehr.) Cleve (Pl. 35, Fig. 14)

Sarode and Kamat, 1984, p 162, pl. 19, f. 434

Frustules 60 µ long, 25 µ broad, elliptic with truncate ends in girdle view; valves lunate with convex dorsal side and concave ventral side with rounded ends; raphe arcuate with dorsally bent central pores; axial area very narrow; central area large, quadrate, reaching the ventral side; striae 12-13 in 10 µ, coarsely punctate, radial on dorsal side.

### 95) *A. veneta* Kuetz. (Pl. 35, Fig. 15)

Sarode and Kamat, 1984, p 163, pl. 19, f. 437

Frustules broadly elliptic with somewhat subtruncate ends in girdle view; valves 15 µ long, 5 µ broad, strongly convex on the dorsal side and slightly concave on the ventral side with inwardly bent rounded ends; raphe thin, very close to ventral margin with central pores dorsally directed; striae 22-24 in 10 µ, clearly punctate, in the middle and very finely punctate, towards the ends.

## Genus: *Cymbella* Ag.

### 96) *Cymbella affinis* Kuetz. (Pl. 36, Fig. 1)

Sarode and Kamat, 1984, p 166, pl. 19, f. 440

Valves 17.5-36 µ long, 8.5-11.5 µ broad, asymmetrical dorsal side strongly convex and ventral side slightly convex; ends constricted, rostrate to subcapitate; raphe slightly thick, axial area narrow, sublinear; central area slightly widened, striae 10-12 in 10 µ, slightly radial, lineate.

### 97) *C. amphicephala* Naeg. (Pl. 36, Fig. 2)

Sarode and Kamat, 1984, p 166, pl. 19, f. 441

Valves 30 µ long 7.5 µ broad, asymmetrically semilanceolate; dorsal side strongly convex; ventral side slightly convex in the middle; ends constricted, produced and capitate; raphe thin and straight with terminal fissures dorsally bent; axial area very narrow; central area small; striae 12-14 in 10 µ, very slightly radial

### 98) *C. aspera* (Ehr.) Cleve (Pl. 36, Fig. 3)

Sarode and Kamat, 1984, p 167, pl. 19, f. 442

Valves 137.5 µ long 27.5 µ broad, asymmetrical with strongly convex dorsal and straight or slightly convex ventral side; ends obtusely rounded; raphe thick, arcuate, slightly excentric with large, ventrally bent central pores and dorsally directed terminal fissures; axial area moderate, linear; striae 8-10 in 10 µ, radial, clear and coarsely punctate.

### 99) *C. bengalensis* Grun. (Pl. 36, Fig. 4)

Sarode and Kamat, 1984, p 167, pl. 19, f. 444

Valves 85 µ long 25 µ broad, asymmetrical, dorsal side strongly convex and ventral side slightly convex with slightly constricted broadly rounded ends; raphe thick with prominent central pores and directed terminal fissures; axial area moderate; central area slightly formed with an isolated stigma on the ventral side; striae 8-10 in 10 µ, radial and coarsely punctate.

### 100) *C. chandolensis* Gandhi (Pl. 36, Fig. 5)

Sarode and Kamat, 1984, p 168, pl. 20, f. 445

Valves 22.5 µ long 5 µ broad, with dorsal margin uniformly convex and ventral margin straight; ends indistinctly constricted, produced, acutely rounded; raphe thin and straight, close to the ventral side, central pores distinct and terminal fissures ventrally directed; axial area very narrow, linear; central area slightly formed towards the dorsal side; striae 16-18 in 10 µ, fine and slightly radial.

### 101) *C. cistula* (Hemp.) Grun. v. *woosungensis* Voigt (Pl. 36, Fig. 6)

Sarode and Kamat, 1984, p 168, pl. 20, f. 446

Valves 95 µm long, 23 µm broad, asymmetrical with dorsal margin strongly convex and ventral margin almost straight or slightly concave and slightly inflated in the middle; ends broadly rounded; raphe thick and straight with central pores curved towards the dorsal side; striae 8-9 in 10 µ, distinctly punctuate and radial.

### 102) *C. gracilis* (Rabh.) Cleve (Pl. 36, Fig. 7)

Sarode and Kamat, 1984, p 170, pl. 20, f 451

Valves 23.5-33 µ long, 4.5-6.5 µ broad, asymmetrical, semilanceolate with dorsal margin uniformly convex in a large arc and ventral margin nearly straight; ends very slightly constricted, raphe thin and straight, close to the ventral margin with central pores, dorsally bent and terminal fissures ventral curved; striae 10-12 in 10 µ, fine, punctate and radial.

### 103) *C. hungarica* (Grun.) Pant. v. *signata* (Pant.) A.Cl. (Pl. 36, Fig. 8)

Sarode and Kamat, 1984, p 170, pl. 20, f. 453

Valves 45 µ long 10 µ broad, asymmetrical with dorsal margin strongly convex and ventral margin more or less straight with an inflation in the middle; ends narrowed, slightly reflexed on the dorsal side and rounded; raphe thick slightly arcuate with central pores and terminal fissures dorsally directed; axial area narrow; central area slightly formed with a stigma on ventral side; striae 8-10 in 10 µ, thick, radial.

### 104) *C. kerkevarensis* A.Cl. (Pl. 36, Fig. 9)

Sarode and Kamat, 1984, p 171, pl. 20, f. 454

Valves 19.5-23 µ long, 6.5-7.5 µ broad, asymmetrical with dorsal margin highly convex and ventral margin slightly convex; end slightly constricted and rostrate; raphe thin, slightly apparently straight, axial area very narrow; central area very small; striae 10-12 in 10 µ, radial throughout and punctate.

### 105) *C. laevis* Naeg. (Pl. 36, Fig. 10)

Sarode and Kamat, 1984, p 171, pl. 20, f. 455

Valves 45 µ long 10 µ broad, nearly lanceolate, dorsal margin convex and ventral margin nearly straight or with a slight median inflation; ends rounded; raphe slightly thick, almost straight; axial area narrow, linear; central area small; striae 12-14 in 10µ, lineate, radial, more widely set on the dorsal side.

### 106) *C. lanceolata* (Ehr.) V. H. (Pl. 36, Fig. 11)

Sarode and Kamat, 1984, p 172, pl. 20, f. 456

Valves 151.2-187 µ long, 29.8-35 µ broad, asymmetrical, dorsal margin convex, more tumid in middle; ventral margin straight with tumidity in the middle, ends obtusely rounded; raphe thick, slightly excentric with large ventrally bent central pores and dorsally directed terminal fissures, axial area wide, linear; central area slightly formed with an arcuate marking on the dorsal side; striae 9-10 in 10 µ, in the middle and 11-12 in 10 µ towards the ends, radial, clear and coarsely punctate.

### 107) *C. lanceolata* (Ehr.) V. H.v. *cornuta* (Ehr.) Grun. (Pl. 36, Fig. 12)

Sarode and Kamat, 1984, p 172, pl. 20, f. 457

Valves 136.8 μ long, 28.8 μ broad, curved, asymmetrical, dorsal margin convex, slightly more convex in the middle, ventral margin concave with a strong inflation in the middle; ends broadly rounded; raphe thick, with dorsally directed central pores and bifurcated terminal fissures; axial area narrow; central area small without any isolated dot; striae 7-9 in 10 μ, clearly punctate, strongly radial throughout.

## 108) *C. maharashtrensis* Sarode and Kamat (Pl. 37, Fig. 1)

Sarode and Kamat, 1984, p 173, pl. 21, f. 483

Valves 20-24 μ long, 3.5-4 μ broad, asymmetrical with more convex dorsal margin and slightly convex ventral margin; end produced and rounded; raphe slightly excentric, thin and straight; axial area narrow, linear; central area small, striae 22-24 in 10 μ, fine and radial.

## 109) *C. osmanabadensis* Sarode and Kamat (Pl. 37, Fig. 2)

Sarode and Kamat, 1984, p 174, pl. 21, f. 484

Valves 20 μ long 5 μ broad, asymmetrical with strongly convex dorsal margin and slightly convex ventral margin; ends slightly constricted and rounded; raphe thick and straight with central pores slightly ventrally bent and terminal fissures dorsally directed; axial area narrow; central area small; striae 12-14 in 10 μ, coarse, slightly radial.

## 110) *C. perpusilla* A.Cl. (Pl. 37, Fig. 3)

Sarode and Kamat, 1984, p 174, pl. 20, f. 461

Valves 15-23 μ long, 3-3.7 μ broad, distinctly asymmetrical, scarcely curved, obtuse ends; dorsal and ventral margins slightly convex; raphae excentric, straight; axial area narrow; central area small, striae 14-16 in 10 μ, slightly radial.

## 111) *C. powaiana* Gandhi (Pl. 37, Fig. 4)

Sarode and Kamat, 1984, p 174, pl. 20, f. 462

Valves 60 μ long 17.5 μ broad, asymmetrical, boat shaped, dorsal margin strongly convex, ventral margin straight with a slight inflation in the middle; ends constricted on the dorsal side, distinctly produced and subtruncate; raphe thick, arcuate, nearly central with central pores distinct and terminal fissures reflexed towards the dorsal side; axial area fairly wide, linear; central area large somewhat roundish without any stigmata; striae 7-8 in 10 μ, coarsely lineate, radial throughout.

## 112) *C. pusilla* Grun. (Pl. 37, Fig. 5)

Sarode and Kamat, 1984, p 175, pl. 20, f. 464

Valves 30.5-35 μ long, 7.8-7.9 μ broad, asymmetrical, dorsal margin smoothly convex, ventral margin straight; end slightly constricted, produced and slightly ventrally curved, raphe thin and straight with terminal fissures bent ventrally; axial area narrow, linear, striae 16-18 in 10 μ, lineate.

### 113) *C. tumida* (Breb.) V.H. (Pl. 37, Fig. 6)

Sarode and Kamat, 1984, p 176, pl. 20, f. 468

Valves 75 μ long 19.5 μ broad, asymmetrical and curved, broadly naviculoid, with rostrate poles, dorsal margin convex and ventral margin straight or slightly convex with median expansion; raphe excentric; axial area narrow; central area large, rounded with ventrally placed prominent isolated dot; striae 9-10 in 10 μ, radial and punctate.

### 114) *C. tumida* (Breb.) V.H. f. *ventricosa* Gandhi (Pl. 37, Fig. 7)

Sarode and Kamat, 1984, p177, pl. 20, f. 469

Valves 108 μ long, 28.8 μ broad, asymmetrical, boat shaped, ventral margin concave with a single median tumidity, dorsal margin strongly convex; ends constricted on the dorsal side, produced with obliquely truncate ends; raphe excentric; axial area narrow; central area large; rounded with a ventrally placed isolated dot; striae 7-9 in 10 μ, clearly punctate and radial.

### 115) *C. turgidula* Grun. (Pl. 37, Fig. 8)

Sarode and Kamat, 1984, p 178, pl. 21, f. 472

Valves 35 μ long 10 μ broad, asymmetrical with strongly convex dorsal margin and almost straight ventral margin; ends rostrate and obtuse; raphe excentric, arcuate; axial area narrow; central area slightly widened with two punctae on the ventral side; striae 9-10 in 10 μ, coarsely but clearly punctate.

### 116) *C. ventricosa* Kuetz. (Pl. 37, Fig. 9)

Sarode and Kamat, 1984, p 178, pl. 21, f. 473

Valves 7.7-35 μ long, 3.6-7.8 μ broad, strongly convex on the dorsal side and straightly or slightly convex on the ventral side with acutely rounded ends; raphe thin and straight, axial area narrow; central area small, striae 10-14 in 10 μ, redial, lineate.

### 117) *C. ventricosa* Kuetz. v. *depressa* Krishn. (Pl. 37, Fig. 10)

Sarode and Kamat, 1984, p 179, pl. 21, f. 475

Valves 28.5-29 μ long, 5.5-8 μ broad, strongly asymmetrical with strongly convex dorsal margin and arcuate ventral margin with a slight median depression; ends obtuse, slightly constricted on the dorsal side; raphe thin and arcuate; axial area somewhat broad; central area slightly dilated; striae 9-10 in 10 μ, distinctly punctate and radial.

### 118) *C. vidarbhensis* Sarode and Kamat (Pl. 37, Fig. 11)

Sarode and Kamat, 1984, p 179, pl. 21, f. 480

Valves 65 µ long 15 µ broad, lunate, dorsal margin strongly convex and slightly depressed towards both the ends; ventral margin almost straight and ventrally gibbous; ends rounded; raphe strongly excentric, straight with terminal fissures ventrally directed; axial area moderate, linear, central area small; striae 9-10 in 10 µ, strong and coarse, radial throughout.

## Family: Bacillariaceae

## Genus: *Nitzschia* Hassall

### 119) *Nitzschia intermedia* Hantzsch (Pl. 38, Fig. 1)

Sarode and Kamat, 1984, p 218, pl. 26, f. 598

Valves 100 µ long, 5 µ broad, linear lanceolate, elongated with cuneate, constricted, produced subcapitate ends; keel punctae 8-10 in 10 µ, small; striae about 25 in 10 µ, fine.

### 120) *N. perminuta* Grun. (Pl. 38, Fig. 2)

Sarode and Kamat, 1984, p 223, pl. 26, f. 614

Valves 25 µ long, 3.5 µ broad, linear lanceolate with constricted capitate ends; keel excentric, keel punctae about 16 in 10 µ, small and round; striae about 18 in 10 µ, punctate.

### 121) *N. obtusa* W. Smith (Pl. 38, Fig. 3)

Sarode and Kamat 1984, p 221, pl. 26, f. 608

Valves 71-143 µ long, 7.5-10.5 µ broad, linear with obliquely cuneate rounded ends in opposite directions; keel strongly excentric, sometime with a slight notch in the middle; keel punctate 8-9 in 10µ, rounded, two of the middle ones widely set; striae about 30 in 10 µ, fine and not clearly visible.

### 122) *N. obtusa* W. Smith v. *scalpelliformis* Grun. (Pl. 38, Fig. 4)

Sarode and Kamat, 1984, p 222, pl. 26, f. 609

Valves 75-129µ long, 5.5-7.5µ broad, linear, very slightly sigmoid with obliquely wedge shaped acutely rounded ends; keel excentric, keel notched in the middle, keel punctate 7-8 in 10 µ, large rounded, striae 30-32 in 10 µ, very fine.

## Sub-order: Surirellineae

## Family: Surirellaceae

## Genus: *Surirella*Turpin

### 123) *Surirella capronii* Breg (Pl. 38, Fig. 5)

Sarode and Kamat, 1984, p 230, pl. 27, f. 641

Valves 150 μ long, 57.5 μ broad, heteropolar, ovate with broadly rounded, somewhat narrowed apex and cuneate acutely rounded base; middle line present, discontinuous, with strong spines developed on elevated cushion; axial area narrowly lanceolate; marginal folds strongly developed, with clear projections, flap windows quite evident; costae 9-13 in 100 μ, thick, radial at the ends; striae fairly visible.

### 124) *S. capronioides* Gandhi (Pl. 38, Fig. 6)

Sarode and Kamat, 1984, p 231, pl. 27, f. 642.

Valves 80-130.6 μ long, 42-50 μ broad, heteropolar, ovate with cuneate base; axial field narrowly lanceolate with a middle line interrupted throughout and with spines at both ends; flap marginal with clear flap projections; costae 24-26 in 100 μ, strong, radial at the ends; striae fine and indistinct.

### 125) *S. ovata* Kuetz. (Pl. 38, Fig. 7)

Sarode and Kamat, 1984, p 232, pl. 28, f. 648, 649

Valves 60 μ long, 20 μ broad, heteropolar, broadly ovate to ovate lanceolate with somewhat narrower base; pseudoraphe narrow, linear; marginal folds scarcely developed; costae 35-60 in 100 μ, thick, rib like, alternating with 2-5 striae; striae 16-20 in 10 μ, very distinct and coarse.

### 126) *S. robusta* Ehr. (Pl. 38, Fig. 8)

Sarode and Kamat, 1984, p 233, pl. 28, f. 652

Valves 113-164 μ long, 51-60 μ broad, heteropolar, broad to narrowly ovate with broadly rounded ends; axial field narrowly lanceolate with median line, marginal folds very strongly developed with thick wavy projections; costae 8-11 in 100 μ, very strong, radial at the ends; striae indistinct.

### 127) *S. tenuissima* Hustedt (Pl. 38, Fig. 9)

Sarode and Kamat, 1984, p 235, pl. 28, f. 657

Valves 23-44.5μ long, 6-8.2μ broad, small, heteropolar, narrowly lanceolate ovate with broadly rounded apex and gradually attenuated acutely cuneate base; axial area indistinct; margin with clear projections; costae about 35-50 in 100 μ, coarse, reaching the middle line.

# Class: Cyanophyceae

## Order : Chroococcales

## Family : Chroococcaceae

## Genus : *Microcystis* Kuetzing

### 1) *Microcystis aeruginosa* Kuetz. (Pl. 39, Fig. 1)

Desikachary, 1959, p 93, pl. 17, f. 1

Colony rounded or slightly longer than broad, solid, when old becoming clathrate, with distinct hyaline colonial mucilage; cells 5 µ in diameter, spherical, generally with gas vacuoles.

### 2) *M. flos-aquae* (Wittr.) Kirchner (Pl. 39, Fig. 2)

Desikachary, 1959, p 94, pl. 17, f 11 and pl 18, f 11

Colonies free-swimming, blue-green, roughly spherical or often squarish in optical section, not clathrate, with in distinct hyaline and homogenous colonial mucilage; cells 5 µ in diameter, spherical, air spaces not present.

### 3) *M. marginata* (Menegh.) Kuetz. (Pl. 39, Fig. 3)

Desikachary, 1959, p 87, pl. 17, f. 5

Colony round or irregularly flattened and more or less lenticular, generally simple; margins of colonial mucilage very distinct, refractive, sometimes stratified; single colony ellipsoidal to ovoid in outline; cells 5 µ in diameter, closely arranged, with gas vacuoles.

### 4) *M. pseudofilamentosa* Crow (Pl. 39, Fig. 4)

Desikachary, 1959, p 94, pl. 18, f. 9 and pl. 20, f. 1

Colonies planktonic, blue-green, very long and narrowing, consisting of a series of parietal colonies and constricted at intervals, reticulate; margins of colonial mucilage indistinct; colonies varying greatly in size, frequently 252 µ in length, 28.3 µ in width; parietal colonies being about equal in length and width; cells 3.7 µ diameter, spherical.

### 5) *M. robusta* (Clark) Nygaard (Pl. 39, Fig. 5)

Desikachary, 1959, p 85, pl. 17, f. 7-10

Colonies planktonic, blue-green, at first round, later irregularly elongate and clathrate; sheath distinct, cells 6.3 µ diameter, without gas-

## Genus: *Chroococcus* Naeg.

### 6) *Chroococcus giganteus* West, W. (Pl. 39, Fig. 6)

Desikachary, 1959, p 101, pl. 26, f. 1

Cells mostly three, seldom 2-4 bright blue-green, without sheath 5 μ broad, with sheath 63μ broad, lamellated, colorless, colonies 14.5 μ in diameter.

### 7) *C. indicus* Zeller (Pl. 39, Fig. 7)

Desikachary, 1959, p 109, pl. 26, f. 10

Thallus gelatinous, thin, a pale brownish; cells single, spherical, 18.9 μ diameter, greenish; sheath hyaline, conspicuous, contents granular.

### 8) *C. limneticus* Lemm. (Pl. 39, Fig. 8)

Desikachary, 1959, p 107, pl. 26, f. 2

Cells spherical, free floating in a gelatinous layer, sheath colorless, colonial mucilage broad, cell contents gray, olive green. Cells 7.5 μ in diameter.

### 9) *C. macrococcus* (Kuetz.) Rabenh. (Pl. 39, Fig. 9)

Desikachary, 1959, p 101, pl. 27, f. 9

Thallus mucilaginous, somewhat broad, yellowish brown, more or less dilated; cells spherical, 2-4 together, also single, 45 μ diam., with sheath 67.5 μ diam.; sheath thick, colorless, lamellated.

### 10) *C. minutus* (Kuetz.) Nag. (Pl. 39, Fig. 10)

Desikachary, 1959, p 103, pl. 24, f. 4

Cells spherical or ablong, single or in groups of 2-4, light blue green, with sheath 15 μ diam. and without sheath 10 μ diam, colonies 17.5 μ; sheath not lamellated, colourless.

### 11) *C. montanus* Hensgirg (Pl. 39, Fig. 11)

Desikachary, 1959, p 108, pl. 26, f. 12

Thallus slimy, gelatinous, brownish, cell 5.6μ diameter, in groups of 2, colonial sheath lamellated, colony 19μ in diameter.

### 12) *C. schizodermaticus* West (Pl. 39, Fig. 12)

Desikachary, 1959, p 103, pl. 26, f. 17

Cells in group of two to form, blue green, without sheath 10 μ, with sheath 30 μ diam; sheath yellow to brown, very distinct, lamellated, outer layers often broken.

### 13) *C. tenax* (Kirchn.) Hieron. (Pl. 39, Fig. 13)

Desikachary, 1959, p 103, pl. 26, f. 7, 16

Cells mostly in groups of two, blue-green colored, without sheath 18.9 μ, with sheath 33.8 μ in diameter; sheath colorless, very thick, very distinctly lamellated, 3 lamellae.

### 14) *C. turgidus* (Kuetz.)Nag. (Pl. 39, Fig. 14)

Desikachary, 1959, p 101, pl. 26, f. 6

Cells spherical or ellipsoidal single, or in groups of mostly 2-4, very seldom many, blue-green, olive green or yellowish, without sheath 30 μ, with sheath 20 μ diam; sheath colourless, not distinctly lamellated.

## Genus: *Gloeocapsa* Kuetz.

### 15) *Gloeocapsa calcarea* Tilden (Pl. 40, Fig. 1)

Desikachary, 1959, p 115, pl. 24, f. 6

Thallus with calcium incrustation; free floating, cells with individual sheath, 7.5 μ diameter, blue green; sheath colorless, often thin; colonies with 3 cells, colonies 51 μ in diameter.

### 16) *G. kuetzingiana* Nag. (Pl. 40, Fig. 2)

Desikachary, 1959, p118, pl. 23, f. 4

Thallus thin, soft, brownish or blackish; cell densely arranged in colony, 125 μ diam.; cells without sheath 5 μ diam.,with sheath 7.5 μ diam., blue green; sheath yellow to brown, not lamellated.

### 17) *G. magma* (Breb.) Kuetz. (Pl. 40, Fig. 3)

Desikachary, 1959, p 120, pl. 24, f. 13

Colony spherical, brown coloured with cuticle like sheath, 37.5 μ in diameter; cells spherical, blue-green and measure 5 μ in diameter.

### 18) *G. punctata* Nag. (Pl. 40, Fig. 4)

Desikachary, 1959, p 115, pl.23, f 2

Thallus gelatinous, light blue green; cells with sheath 5 μ broad, cell without sheath 1.5 μ, blue-green; sheath thick, colourless, unlamelleated or scarcely lamellated; cells 2-16 in groups or colonies, about 25 μ diam.

## Genus: *Aphanocapsa* Naeg.

### 19) *Aphanocapsa biformis* A.Br. (Pl. 40, Fig. 5)

Desikachary, 1959, p 134, pl. 21, f. 3

Thallus olive green, gelatinous, often expanding; cells 5 µ diam, spherical, mostly with a special envelope; loosely arranged, 2-4 together in a common mucilaginous envelope, nannocytes about 2 µ diam.

**20) *A. grevillei* (Hass.) Rabenh. (Pl. 40, Fig. 6)**

Prasad and Srivastava, 1992, p 36, pl. 5, f. 2

Thallus light blue green, spherical densely arranged and embedded in gelatinous matrix; cells spherical, 2.5 µ in diameter, closely arranged, individual envelope not distinct; cell contents blue green, homogenous without gas vacuoles; cell wall thin and smooth

### Genus: *Merismopedia* Meyen

**21) *Merismopedia elegans* A.Br. (Pl. 40, Fig. 7)**

Prescott, 1951, p 459, pl. 101, f. 1

Colony irregularly quadrangular, composed of 16 or more compactly arranged, ovate cells, with the row of cells, becoming distorted in older and longer colonies; cells 5 µ in diameter, 7.5 µ long; contents bright blue green.

**22) *M. glauca* (Ehr.) Nag. (Pl. 40, Fig. 8)**

Desikachary, 1959, p 155, pl. 29, f. 5

Colonies planktonic, mostly small with 64 cells, rarely more, 137.5 µm in diameter, cells oval, closely arranged, 5µ broad, pale blue green.

**23) *M. minima* Beck (Pl. 40, Fig. 9)**

Desikachary, 1959, p 154, pl 29, f 11

Cells pale blue green, 4 to many in small colonies, 0.5 µ broad, free swimming, groups of four cells 2.5 µ.

**24) *M. punctata* Meyen (Pl. 40, Fig. 10)**

Prescott, 1951, p 459, pl. 102, f. 10

A rectangular plate of 64 ovate cells, usually loosely arranged, sometimes in compact groups of 4-8 individuals, the groups are widely seprated within a broad gelatinous envelope, cells 2.5 µ.

### Genus: *Gloeothece* Nag.

**25) *Gloeothece rhodochlamys* Skuja (Pl. 40, Fig. 11)**

Desikachary, 1959, p 125, pl. 25, f. 13

Thallus minute and microscopic, more rounded, blue green, red or purple; cells ellipsoidal to short cylindrical with rounded ends, 2.5 µ in size.

### Genus: *Gomphosphaeria* Kutz.

**26) *Gomphosphaeria aponina* Kuetz. (Pl. 40, Fig. 12)**

Desikachary, 1959, p 150, pl. 28, f. 3

Cells pyriform or cuneate, cordate at longitudinal cell division, 12.5 μ broad and 15 μ long, blue-green or yellowish mostly with a distinct mucilaginous envelope, placed at the ends of regularly dichotomously branched radial mucilaginous stalks; colony large 62.5 μ in diameter.

### Order: Chamaesiphonales

### Family: Dermocarpaceae

### Genus: *Stichosiphon* Geitler

**27) *Stichosiphon sansibaricus* (Hieron.)Drouet et. Daily (Pl. 40, Fig. 13)**

Desikachary, 1959, p 176, pl. 32, f.12

Plants microscopic, blue green, olivaceous, solitary, erect or curved, epiphytic, 7.5 μ broad, 200 μ high; contents dividing into a series of spherical, ovoid endospores, endospores generally uniseriate; sheath hyaline, thin somewhat stalked at the base and attached; contents blue green, olivaceous, or rose, homogenous.

### Order: Nostocales

### Family: Oscillatoriaceae

### Genus: *Arthrospira* Stizenberger

**28) *Arthrospira platensis* (Nordst) Gomont f. *granulata* Desikachary (Pl. 41, Fig. 1)**

Desikachary, 1959, p 190, pl. 35, f. 5

Trichome light blue-green, not constricted at the cross walls, 7.5 μ broad, slightly less broad at the ends, spirally coiled, spirals at the ends being narrower than the middle, spirals 30 μ and distinct; cells discoid, cross walls often indistinct sometimes distinct, granulated.

**29) *A. platensis* (Nordst) Gomont. v. *non-constricta* (Banerji)**

Desikachary (Pl. 41, Fig. 2)

Desikachary, 1959, p 190-191, pl. 36, f. 1, 2

Thallus blue-green, free-floating, trichomes without sheath, slightly constricted at the cross-walls, not attenuated at the ends. Less regularly spiraley coiled; ends cells broadly rounded. Trichomes 6.3 μ broad, unconstricted, cells 3.2μ long, spirals 37.5 μ broad and 52.3 μ distant.

## Genus: *Spirulina* Turpin em. Gardner

### 30) *Spirulina gigantea* Schmidle (Pl. 41, Fig. 3)

Desikachary, 1959, p197, pl. 36, f. 14

Trichomes 5 μ broad, blue green, regularly spirally coiled, at the end conical attenuated, spirals 15 μ broad.

### 31) *S. major* Kuetz. (Pl. 41, Fig. 4)

Prescott, 1951, p 480, pl. 108, f. 11

Trichomes loosely spiralled, scattered among other algae, or when aggregated forming a dark, blue green mass; trichomes 1.5 μ in diameter; spirals 2.5 μ wide; distance between spirals 2.5 μ.

### 32) *S. meneghiniana* Zanard. ex Gomont (Pl. 41, Fig. 5)

Desikachary, 1959, p 195, pl. 36, f. 8

Trichome without sheath, free-floating, 3.0 μ broad, flexible, irregularly spirally coiled, bright blue-green, forming a thick blue-green thallus; spirals 5 μ broad and 3.2 μ distant from each other.

### 33) *S. princeps* W.et.G.S. West (Pl. 41, Fig. 6)

Desikachary, 1959, p 197, pl. 36, f. 7

Trichome 5 μ broad, short, blue green, regularly spirally coiled, spirals 10 μ broad and 10 μ distant.

## Genus: *Oscillatoria* Vaucher

### 34) *Oscillatoria acuta* Bruhl *et* Biswas, orth. Mut. Geitler (Pl. 41, Fig. 7)

Desikachary, 1959, p 240, pl. 39, f. 5

Trichomes solitary, hardy, brittle not constricted at the cross walls; 90 μ long, usually quite straight, narrow at the apex, which may be straight but is more often abruptly bent aside; cells 3 μ long, contents bluish green, finely granular, sometimes with some larger granules close to the surface.

### 35) *O. amphibia* Ag. ex. Gomont (Pl. 41, Fig. 8)

Desikachary, 1959, p 229, pl. 37, f. 6

Thallus deep blue-green; planktonic, trichome straight or coiled, apices not attenuated, not capitate, not tapering towards the apex, not constricted at the cross-walls, 4.2 µ broad, cells 2-3 times longer than broad, 8 µ long, end cell not capitate, rounded, calyptra absent.

### 36) *O. annae* van Goor (Pl. 41, Fig. 9)

Desikachary, 1959, p 203, pl. 38, f. 13

Trichome straight, dull blue-green, constricted at the cross-wall, 12µ broad, mostly attenuated at the ends, and bent; cells 1/3-15 as long as broad, 3.7µ long, not granulated at the cross-walls; end-cell rounded, calyptra absent.

### 37) *O. biswasii* Kamat (Pl. 41, Fig. 10)

Kamat, 1963, p 282, pl 14, f 80

Plant mass blue-green; trichomes straight or bent, tapering towards the apices, constricted at the cross-walls; cells shorter than broad, 8.4 µ in diameter, 8.5 µ long; end cell conical without cap or calyptra.

### 38) *O. chlorina* Kuetz. ex Gomont (Pl. 41, Fig. 11)

Desikachary, 1959, p 215, pl. 40, f. 4

Thallus very thin, green, trichome straight or curved, unconstructed at the cross walls; gas vacuoles absent; cells somewhat longer than broad, 7.5 µ long, cross walls not granulated; calyptra absent.

### 39) *O. cortiana* Meneghini ex. Gomont (Pl. 41, Fig. 12)

Desikachary, 1959, p 233, pl. 38, f. 14

Thallus dull blue green; trichome straight, slightly constricted at the joints, gradually tapering at the ends, bent, not capitate, cells as long as broad, 7.5 µ long, septa not granulated; end cells abtuse, without calyptra.

### 40) *O. cortiana* Meneghini ex Gomont v. *minor* Kamat (Pl. 41, Fig. 13)

Kamat, 1963, p 284, pl 16, f 90

Plant mass blue-green; trichomes tapering towards the apices, slightly bent at apices, constricted at the cross-walls; cells longer than broad, 6.2µ in diameter, 7.8µ long, end cell conical, without calyptra or cap.

### 41) *O. curviceps* Ag. ex Gomont (Pl. 42, Fig. 1)

Desikachary, 1959, p 209, pl. 38, f. 2

Thallus light green; trichomes straight, bent at ends; very little attenuated; 15 µ broad; cells not constricted at cross walls, sometimes granulated; end cells flat rounded, not capitate.

### 42) *O. foreaui* Fremy (Pl. 42, Fig. 2)

Desikachary, 1959, p 219, pl. 40, f 18

Trichomes sparse, elongate, bent, apex gently curved, distinctly constricted at the cross-wall, and 5.6 μ broad, apex not attenuated not capitate; cells ½ as long as broad, apical cell rounded, calyptra absent.

### 43) *O. formosa* Bory ex. Gomont (Pl. 42, Fig. 3)

Prasad and Srivastava, 1992, p 62, pl. 9, f. 5

Trichomes elongated, solitary or aggregated to form dark blue green masses almost straight, very gently attenuated towards slightly bent apices; cells quadrate or up to half times as long as broad,5 μ long; cell contents bright blue green, homogenous without granules and gas vacuoles; end cell obtuse or bluntly rounded, non capitate, calyptra absent; cell wall thick, smooth.

### 44) *O. irrigua* (Kutez.) Gomont (Pl. 42, Fig. 4)

Desikachary, 1959, p 224, pl. 42, f. 7, 9

Thallus blackish blue-green, straight, flexuous, not torulose, 6.3 μ broad, apex, slightly attenuated, subcapitate, straight; cells about ½ as long as broad, 3.2μ long, septa ordinarily not granulated; apical cell convex, with an evident thickened outer wall, calyptra present.

### 45) *O. limosa* Ag. ex Gomont (Pl. 42, Fig. 5)

Desikachary, 1959, p 206, pl. 42, f. 11

Thallus dark blue-green, trichome straight, not constricted at cross walls, commonly 15 μ broad; cells 1/3 to 1/6 as long as broad, 5 μ long, cross walls frequently granulated; end cells flatly rounded with slightly thickened membrane.

### 46) *O. margaritifera* (Kuetz.) Gomont (Pl. 42, Fig. 6)

Desikachary, 1959, p 202, pl. 42, f. 8

Trichome olive-green, forming blackish thallus, 15.4μ broad, fragile, straight, constricted at the cross-walls, apices slightly bent, slightly attenuated; cells 1/3-1/7 as long as broad, 6.3 μ long, cross-walls granulated, end cell capitate with slightly convex calyptra.

### 47) *O. ornata* Kuetz. ex Gomont (Pl. 42, Fig. 7)

Desikachary, 1959, p 206, pl. 40, f. 3

Thallus dark blue-green, slightly bent at apices, constricted at the cross-walls, 8 μ broad, dull blue-green, cells 1/2-1/6 as long as broad, 3.7 μ long, cross-walls granulated; apices slightly attenuated; end-cells convex, not capitate, without thickened membrane.

## 48) *O. ornata* v. *crassa* Rao C.B. (Pl. 42, Fig. 8)

Desikachary, 1959, p 206, pl 39, f 11 and pl 41, f 12

Thallus dark blue-green; trichome straight of uniform thickness, 8 μ broad, constricted at the cross-walls, cross-walls granulated; cells shorter than broad, 5 μ long; end-cell convex without calyptra, not captitate.

## 49) *O. princeps* Vaucher (Pl. 42, Fig. 9)

Prescott, 1951, p 489, pl. 110, f. 1

Trichomes solitary or loosely entangled to form small floating plant masses, which are black-green, brownish, or violet in color; trichomes very slightly and briefly tapering at the apex. Apical cell usually not capitate, sometimes very slightly so, the outer membrane broadly convex and smooth. Cells 23 μ in diameter 5.5 μ long; not constricted at the cross walls, which are not granular; cell contents densely granular.

## 50) *O. proboscidea* Gomont (Pl. 42, Fig. 10)

Prasad and Srivastava, 1992, p 69, pl. 8, f. 9

Trichomes single, dark blue green; not constricted at cross walls, 6-12 μ in diameter, cells¼ times as long as broad, 2.5 μ long; cell contents light blue green, homogenous, without gas vacuoles, granules distributed irregularly throughout but usually towards septa; end cells almost truncate, capitate with thick convex outer membrane; cell wall thick and smooth.

## 51) *O. rubescens* D.C. ex Gomont v. *curvata* Kamat (Pl. 42, Fig. 11)

Kamat, 1963, p 289, pl 16, f 94

Plant mass blue-green, sometimes forming a purple red, free swimming bundles, trichome straight, slightly curved at the apex, ends gradually attenuated, 7.5 μ broad, not constricted at the cross-walls, cells 1/2-1/3 as long as broad, 2.5 μ long, often granulated at the septa, end cell capitate with convex calyptra.

## 52) *O. rubescens* D.C. ex Gomont v. *kolhapurensis* Kamat (Pl. 42, Fig. 12)

Kamat, 1963, p 289, pl 16, f 95

Plant mass blue-green, trichomes long, usually straight, tapering towards the apices, not constricted, at cross-walls, cells shorter than broad, 7 μ in diameter, 2.6 μ long; end cell capitate with calyptra. The trichomes are broader than the type specimen.

## 53) *O. sancta* (Kuetz.) Gomont (Pl. 42, Fig. 13)

Desikachary, 1959, p 203, pl. 42, f. 10

Trichomes aggregated to form a dark gray-green plant mass, usually on submerged vegetation; straight, not tapering towards the apex, distinctly constricted. Apical cell slightly capitate, with a calyptra, and with a thickened outer membrane. Cells 8.4 μ in diameter, 3.5 μ long; slightly constricted at the cross-walls, granular.

### 54) *O. subbrevis* Schmidle (Pl. 42, Fig. 14)

Prasad and Khanna, 1987, p 256, f 1

Thallus pale gray-green. Trichomes single, 10.5μ broad, nearly straight, not tapering towards the apices, not attenuated at the apices; cell contents granular, cell 2.5 μ long, not constricted at the cross wall; end cell rounded, calyptra absent.

## Genus: *Lyngbya* Ag.

### 55) *Lyngbya antartica* Gain v. *maior* Kamat (Pl. 43, Fig. 1)

Kamat, 1963, p 294, pl 17, f 99

Filaments usually solitary, blue-green; 15.3 μ diameter; sheath thin, yellow brown; trichomes attenuated at the apices, not constricted at the cross-walls, cells shorter than broad, 11 μ in diameter, 1.7 μ long; end cell capitate without calyptra.

### 56) *Lyngbya birgei* Smith, G.M. (Pl. 43, Fig. 2)

Desikachary, 1959, p 296, pl. 50, f. 8

Filaments straight, seldom, coiled, free floating, 22.5 μ broad, sheath firm, colourless, seldom lamellated 2.5 μ thick; trichome not constricted at the cross walls, ends rounded, not attenuated, not capitate; cells 2.5 μ long

### 57) *L. confervoides* C.Ag. ex. Gomont (Pl. 43, Fig. 3)

Desikachary, 1959, p 314, pl. 49, f. 9 and pl 52, f 13

Thallus blue-green, filament straight; sheath colourless, not constricted at the cross walls, not attenuated at the apices 8 μ broad; cells 1/3-1/8 times as long as broad, 2.5 μ long; end cell round, calyptra absent.

### 58) *L. connectens* Bruhl et Biswas (Pl. 43, Fig. 4)

Desikachary, 1959, p 308, pl. 51, f. 5,6,11

Stratum extensive, lying parallel to each other; filaments straight, sheath firm, brownish, dark green; trichomes 7 μ broad, not constricted at the cross wall, cells about 1/6 as long as broad, 1.2 μ long.

### 59) *L. hieronymusii* Lemm. (Pl. 43, Fig. 5)

Desikachary, 1959, p 297, pl. 48, f. 4

Filament single, free-floating, straight, 13.8μ broad; sheath firm, homogenous, colourless, cells 12.6μ broad, 3.7μ long, not constricted at the cross-walls, granulated, end cell broadly rounded.

### 60) *L. latissima* Prescott (Pl. 43, Fig. 6)

Prescott, 1951, p 501, pl. 112, f. 9

Plants solitary, planktonic, entangled among other floating algae; trichome straight, not tapering towards the apices; cells disc like, with contents finely and evenly granular, 37.5 μ in diameter, 7.5 μ long; sheaths thick, 5 μ wide, lamellated, with outer layers wrinkled and roughned in age; filaments 50 μ in diameter.

### 61) *L. maharashtrensis* Kamat (Pl. 43, Fig. 7)

Kamat, 1963, p 299, pl 17, f 103

Plant-mass blue-green; filament 12.6 μ broad, usually straight, sheath thin, unstratified, hyaline to yellow-brown; trichomes not tapering towards the apices, not constricted at the cross-walls; cells much shorter than broad, 1/4-1/3 times as long as broad, 14 μ diameter, 2.5 μ long; end cell rounded without calyptra.

### 62) *L. major* Meneghini (Pl. 43, Fig. 8)

Prescott, 1951, p 502, pl. 112, f. 10

Plant mass dark green; entangled with other algae, filament straight or slightly curved, somewhat gregarious, sheath thick, stratified, yellow brown; trichomes not constricted and not granulated at the cross-walls, not attenuated at the ends; cells much shorter than broad, 24 μ in diameter, 2.8 μ long; end cell rounded without calyptra.

### 63) *L. majuscula* Harvey ex Gomont (Pl. 43, Fig. 9)

Desikachary,1959, p 313, pl.48, f.7, pl.49, f.12, and pl.52,f.10

Thallus expanded, dull blue-green, filament very long, sheath colourless, lamellated up to 11.2 μ thick, not constricted at the cross-walls, not attenuated at the ends, 25.2 μ broad, cells very short 1/5-1/4 times as long as broad, 4.1 μ long, end cells rounded, calyptra absent.

### 64) *L. martensiana* Menegh. ex. Gomont. (Pl. 43, Fig. 10)

Desikachary, 1959, p 318, pl. 52, f. 6

Thallus caespitose, blue green; sheath colourless,thick; trichome 10 μ broad, not constricted at cross walls, cross wall sometimes granulated, apices not attenuated, cell 10 μ broad, 2.5 μ in length; end cell rotund,without calyptra.

### 65) *L. rubida* Fremy (Pl. 43, Fig. 11)

Desikachary, 1959, p 298, pl. 53, f. 10

Thallus floccose, less expanded, brownish purple; sheath reddish, filament straight, not constricted at the cross-walls, 5.6μ broad, ends not attenuated, not capitate; 4.8μ long; end cells rounded, calyptra absent.

### 66) *L. stagina* Kuetz. f. *non-granulata* Kamat (Pl. 43, Fig. 12)

Kamat, 1963, p 301, pl 17, f 102.

Filament long, curved, entangled, 16.8 μ broad; sheath thick, stratified, yellow-brown; trichomes not tapering towards the apices, not constricted and not granulated at the cross-walls; cells much shorter than broad, 13.4 μ in diameter, 3.5 μ long; end cell rounded without calyptra.

## Family: Nostocaceae Kuetz.

## Genus: *Cylindrospermum*

### 67) *Cylindrospermum sphaerica* Prasad, B.N. (Pl. 44, Fig. 1)

Desikachary, 1959, p 363, pl. 64, f. 7

Plant mass soft, mucilaginous, pale brown and forming a mat; trichome single, often curved, often entangled with each other, cells barrel shaped, 5 μ long, constricted at the septa; heterocysts ellipsoidal, rounded at distal ends, one at each end of the trichome, 5 μ broad and 7.5 μ long.

## Genus: *Anabaena* Bory

### 68) *Anabaena iyengarii* Bharadwaja (Pl. 44, and Fig. 2)

Desikachary, 1959, p 406, pl. 78, f. 2.

Trichome without sheath, blue-green, uniformly broad single 9.8μ broad, end cell conical with rounded apex; cell barrel-shaped, shorter than broad, heterocysts intercalary nearly spherical, 10.5 μ broad and 8.4 μ long.

### 69) *Anabaena subcylindrica* Borge (Pl. 44, Fig. 3)

Prescott, 1951, p 518, pl. 118, f.7

Trichomes straight; solitary or forming a thin layer; adherent on submerged aquatics and entangled among other algae. Cells short-cylindric, 5 μ in diameter, 7.5 μ long. Heterocyst cylindrical, 7.5 μ broad, 15 μ long. Gonidia cylindrical.

Thallus mucilaginous, mostly dull blue-green; trichome uniformly broad, short, without sheath, but in a common mostly very delicate and often imperceptible, mucilage of thin consistency; cells cylindrical, constricted at the cross-walls; heterocysts terminal, at both ends or at one end only, sometimes intercalary; spores single, rarely in series, next to the heterocyst on one side, much bigger than the vegetative cells.

## Family: Scytonemataceae

## Genus: *Scytonema* Ag.

### 70) *Scytonema arcangelii* Born. *et* Flah. (Pl. 44, Fig. 4)

Prasad and Srivastava, 1992, p 124, pl. 14, f. 1-3

Filaments long, brownish green, compactly interwoven forming expanded floccose flushy thallus; 17.5 µ in diameter, sheaths thick, membranaceous, colourless, smooth, sometimes gelatinised at the points from where branches arise; cells somewhat quadrate or muchshorter than broad, 5 µm long; cell contents olive blue green, homogenous without granules and gas vacuoles; end cells rounded; heterocysts intercalary, cylindrical with flattened ends, bipolar 10 µm broad, 15 µm long; cell wall smooth and thick.

### 71) *S. bohneri* Schmidle (Pl. 44, Fig. 5)

Prasad and Srivastava, 1992, p 125, pl. 15, f. 3

Filaments long, densely entangled, dark blue-green, forming caespitose free floating masses, 15 µ in diameter, branched, branches single or geminate, slightly narrow than main filament, 10 µ broad; sheaths thick, colourless, smooth lamellated 2.5 µ thick; trichome bluish green, not constricted at the cross walls; cells rectangular; cell contents blue green, homogenous, without gas vacuoles, granules present distributed irregularly throughout the cell; end cells rounded heterocysts intercalary, ellipsoidal to cylindrical, compressed with rounded or flattened ends, bipolar, 20 µ long, 10 µbroad, cell wall smooth and thick.

### 72) *S. javanicum* (Kuetz.) Born. ex Born.et. Flah. (Pl. 44, Fig. 6)

Prasad and Srivastava, 1992, p 129, pl. 19, f. 3

Filaments long, flexuous, brownish blue green, densly aggregated into cushion shaped floating mass, 17.5 µ in diameter; sheaths thick, smooth, gelatinous, lamellated, 2.5 µ thick; trichomes blue green; cells composed of quadrate or shorter than broad, 7.5 µ long; cell contents light blue green, homogenous without granules and gas vacuoles end cells rounded; heterocysts yellow, cylindrical, intercalary, bipolar, single 27.5 µ long, 15 µ broad; cell wall smooth and thick.

### 73) *S. ocellatum* Lyngbye ex. Born. *et* Flah v. *capitatum* Ghose (Pl. 44, Fig. 7)

Prasad and Srivastava, 1992, p 135, pl. 16, f. 1

Filaments dark blue green, compactly entangled, forming free floating cushion shaped mass, 15 µ in diameter, branched, pseudobranches short rare, usually single; sheaths thick, hyaline, olive yellow, transparent 2.5 µ thick, lamellated when old; trichome olive blue green, swollen at apex, not constricted at thecross walls; cells quadrate or shorter than broad, 7.5 µ long, cellcontents olive green, homogenous without granules and gasvacuoles; end cells rounded; heterocysts intercalary, usually single,bipolar, ellipsoidal or cylindrical with rounded or flattened ends, 10 µ broad, 20 µ long; cell wall mooth, thick.

### 74) *S. simplex* Bharadwaja (Pl. 44, Fig. 8)

Prasad and Srivastava, 1992, p 131, pl. 17, f. 4

Filaments long, irrigularly curved, loosely entangled forming pale blue green, thick thallus, sparingly branched, pseudobranches usually geminate overlaping to each other 15 µ in diameter; sheathsthick, transparent yellowish brown, unstratified becoming lamellated at maturity; trichomes 8-9.5 µ broad, olive blue green, neither attenuated at apices nor constricted at the cross walls; cells in apical portions usually quadrate but in rest of the portions elongated, cylindrical, about two times as long as broad, 17.5 µ long; cell contents blue green, homogenous, without granules and gas vacuoles; end cells broadly rounded; heterocysts intercalary, single, bipolar, elongated cylindrical with rounded ends, 10 µm broad, 32.5 µ long; cell wall smooth and thick.

## Genus: *Plectonema* Thuret.

### 75) *Plectonema radiosum* (Schiederm.) Gomont (Pl. 44, Fig. 9)

Prasad and Srivastava, 1992, p 123, pl. 14, f. 5

Filaments densely interwoven forming dark blue green, caespitose cushion like thallus, elongated flexus, 12.5 µ in diameter; sheaths in lower part thick, hyaline, smooth, colourless, lamellated but in upper parts thin 2.5 µ thick; trichomes blue green, frequently branched, branching both single and geminate, not attenuated, in lower parts not constricted but in upper parts distinctly constricted at the cross walls; cells somewhat quadrate or shorter than broad, 7.5 µ long; cell contents blue green, homogenous, gas vacuoles absent, granules present but distributed towards septa; hormogones not seen; cell wall smooth and thick

## Family: Rivulariaceae

## Genus: *Gloeotrichia* Ag.

### 76) *Gloeotrichia raciborskii* Woloszynska v. *conica* Dixit (Pl. 44, Fig. 10)

Desikachary, 1959, p 563, pl. 117, f. 10

Thallus spherical, soft; sheath at the base, dull brown; cells at the base of trichome shorter than broad; heterocysts spherical, 6 µ broad; spores long, ellipsoidal, with yellowish epispores.

## Order: Stigonematales

## Family: Stigonemataceae

## Genus: *Stigonema* Ag.

### 77) *Stigonema informe* Kuetz. ex Born. *et* Flah. (Pl. 44, Fig. 11)

Desikachary, 1959, p 613, pl. 137, f. 2

Thallus expanded, crustaceous or sometimes occurring as single filaments, brownish or blackish; filaments 70 µ broad, prostrate below and erect above, irregularly branched, branches straight or flexuous, with secondry branches mostly arising from the upper surface, which form hormogones; trichome with 4-6 rows of cells or seldom with many rows; cells 15 µ broad.

## Order: Pleurocapsales

## Family: Pleurocapsaceae

## Genus: *Myxosarcina* Printz.

### 78) *Myxosarcina burmensis* Skuja (Pl. 44, Fig. 12)

Desikachary, 1959, p 178, pl. 32, f. 21

Plants aquatic, minute, microscopic, rounded, sarcinoid; cells more or less angular or with rounded corners, arranged in transverse and vertical series, 2.5 µ in diameter; pale blue green.

### 79) *M. spectabilis* Geitler (Pl. 44, Fig. 13)

Desikachary, 1959, p 178, pl. 30, f. 4

Cells in three dimensional colonies, colonial sheath thin, distinct, and hyaline, individual sheaths occasionally present; cell contents blue green. Cells 7.5 µ broad.

## Class : Euglenophyceae

## Order : Euglenales

## Family: Euglenaceae

## Genus : *Phacus* Dujardin

### 1) *Phacus curvicauda* Swirenko (Pl. 45, Fig. 1)

Prescott, 1951, p 399, pl. 88, f. 21

Cells broadly ovoid to suborbicular in outline, slightly spiral in the posterior part which is extended into caudus; anterior end broadly rounded; periplast longitudinally finely striated; paramylon bodies 2 large discs; chloroplast numerous ovoid bodies; cells 25 μ in diameter, 30 μ long.

### 2) *P. pseudoswirenkoi* Prescott (Pl. 45, Fig. 2)

Prescott, 1951, p 402, pl. 88, f. 14

Cells orbicular in outline, abruptly narrowed posteriorly and produced to form a short, sharp caudus; anterior end broadly rounded; flagellulm about as long as body; periplast longitudinally striated; paramylon body a large, circular plate; cells 30 μ broad and 37.5 μ long.

# Plates

**Map 1: Nathsagar Reservoir, Paithan, Showing the Study Area (Source: Andhale, 2008). (Map not the scale) (Source: Andhale, 2009)**

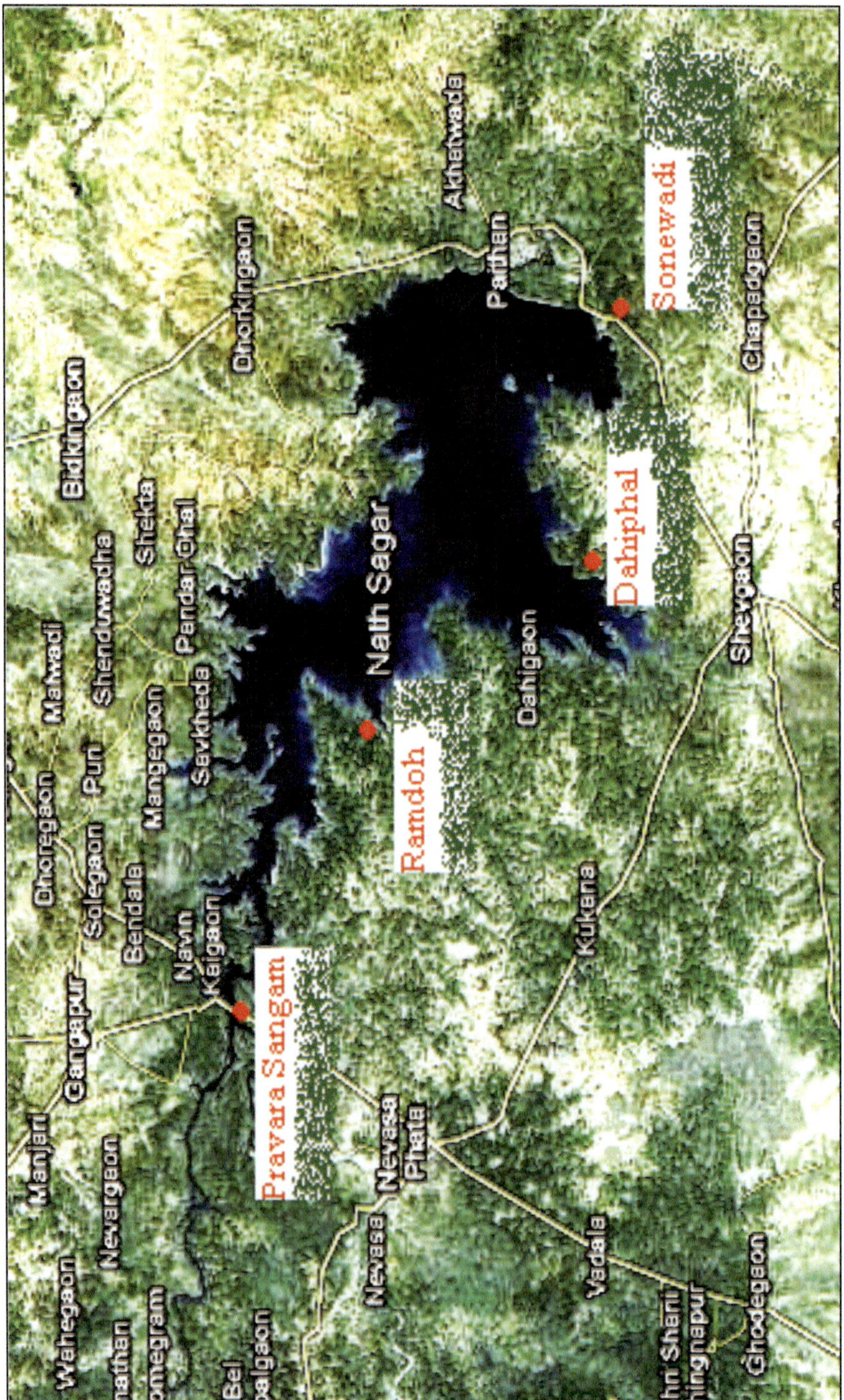


**Map 2: Nathsagar Reservoir, Paithan,Showing the Study Area. (Source: Andhale, 2009) (Source- http://maps.google.co.in/maps?hl=en and tab=wl)**

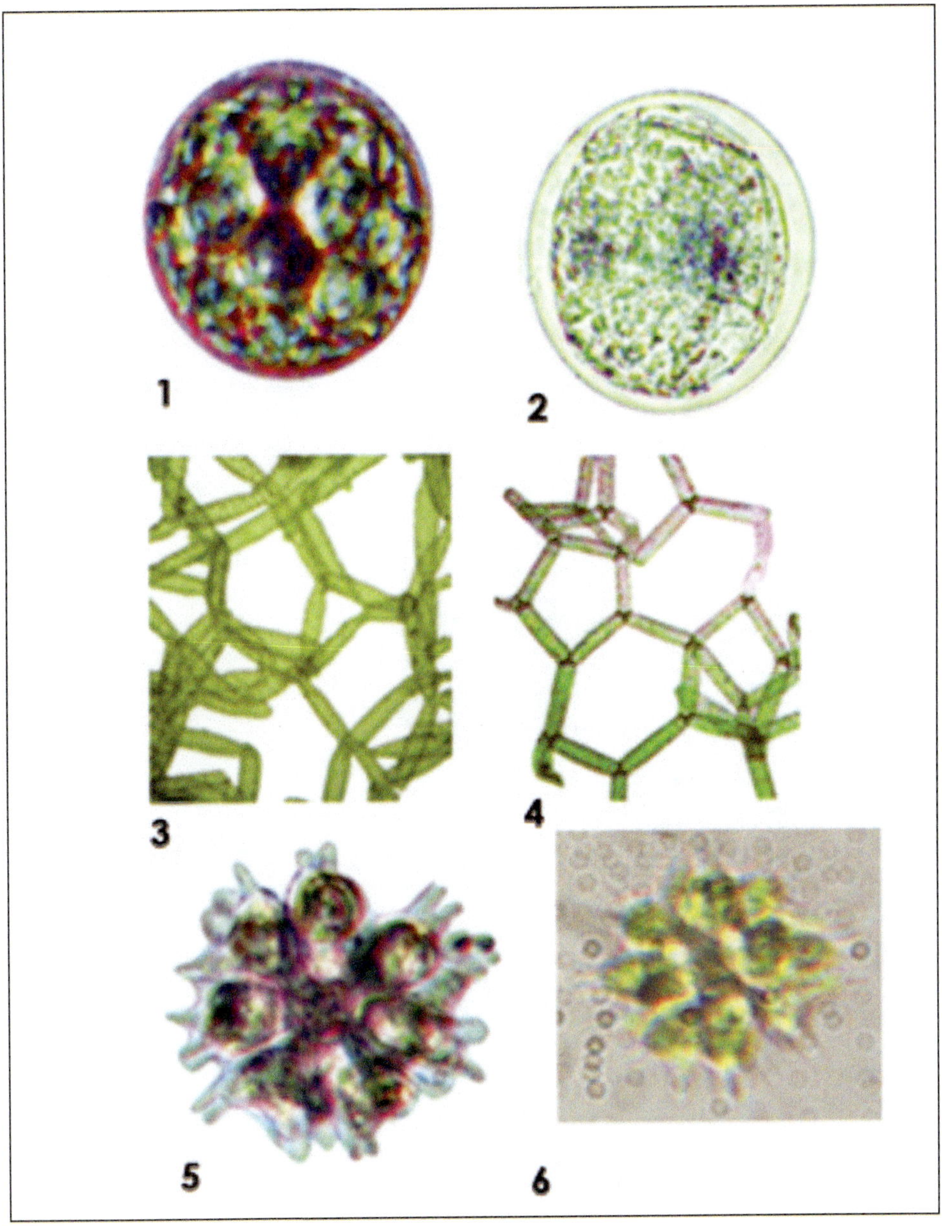



**Plate 1**

**Fig. (1) *Chlorococcum humicola* (2) *C. infusionum* (3) *Hydrodictyon indicum* 4) *H. reticulatum* (5) *Sorastrum americanum* v. *undulatum* (6) *S. spinulosum***

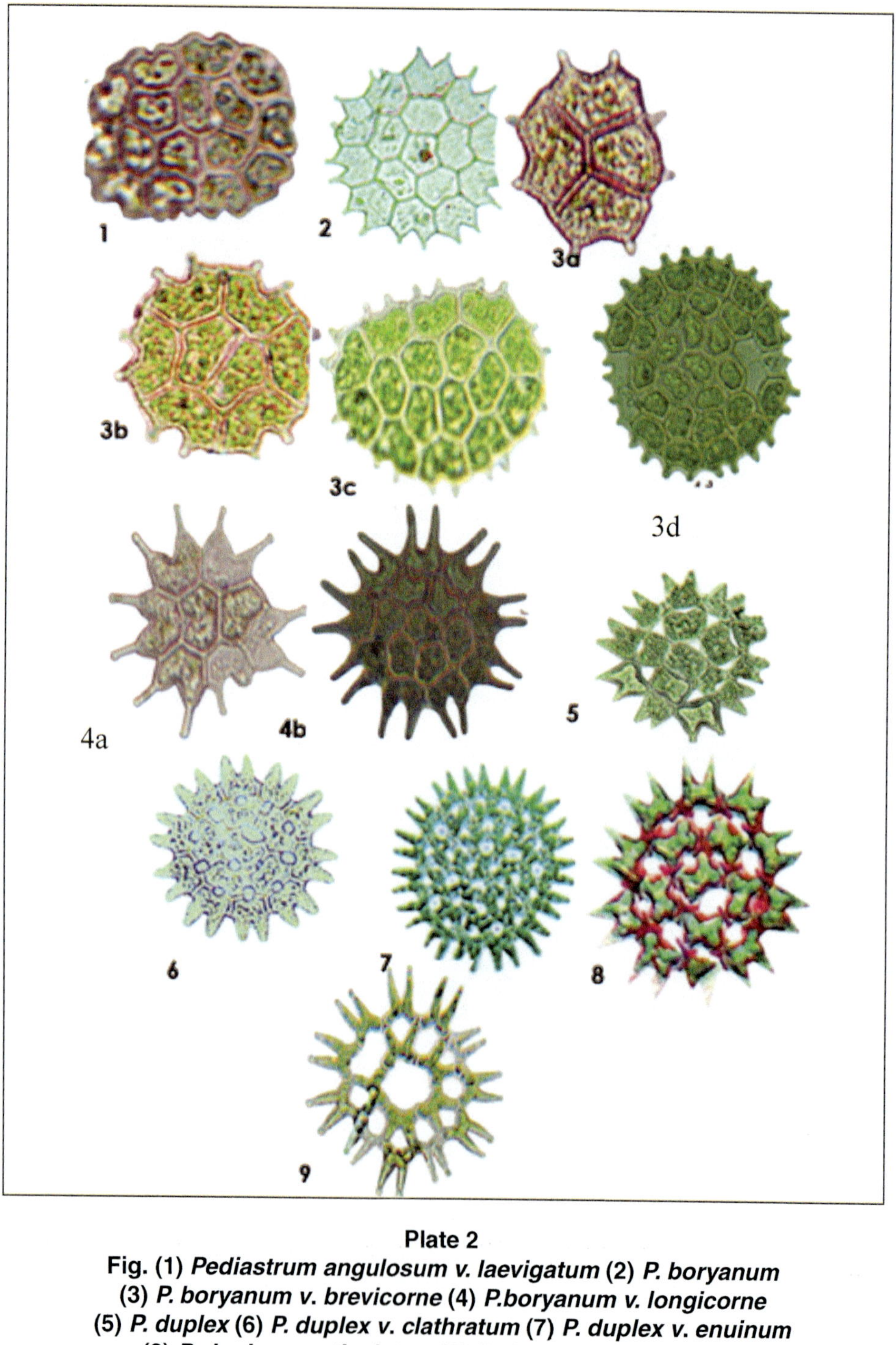


**Plate 2**

**Fig. (1) *Pediastrum angulosum v. laevigatum* (2) *P. boryanum* (3) *P. boryanum v. brevicorne* (4) *P.boryanum v. longicorne* (5) *P. duplex* (6) *P. duplex v. clathratum* (7) *P. duplex v. enuinum* (8) *P. duplex v. reticulatum* (9) *P. duplex v. gracillimum.***

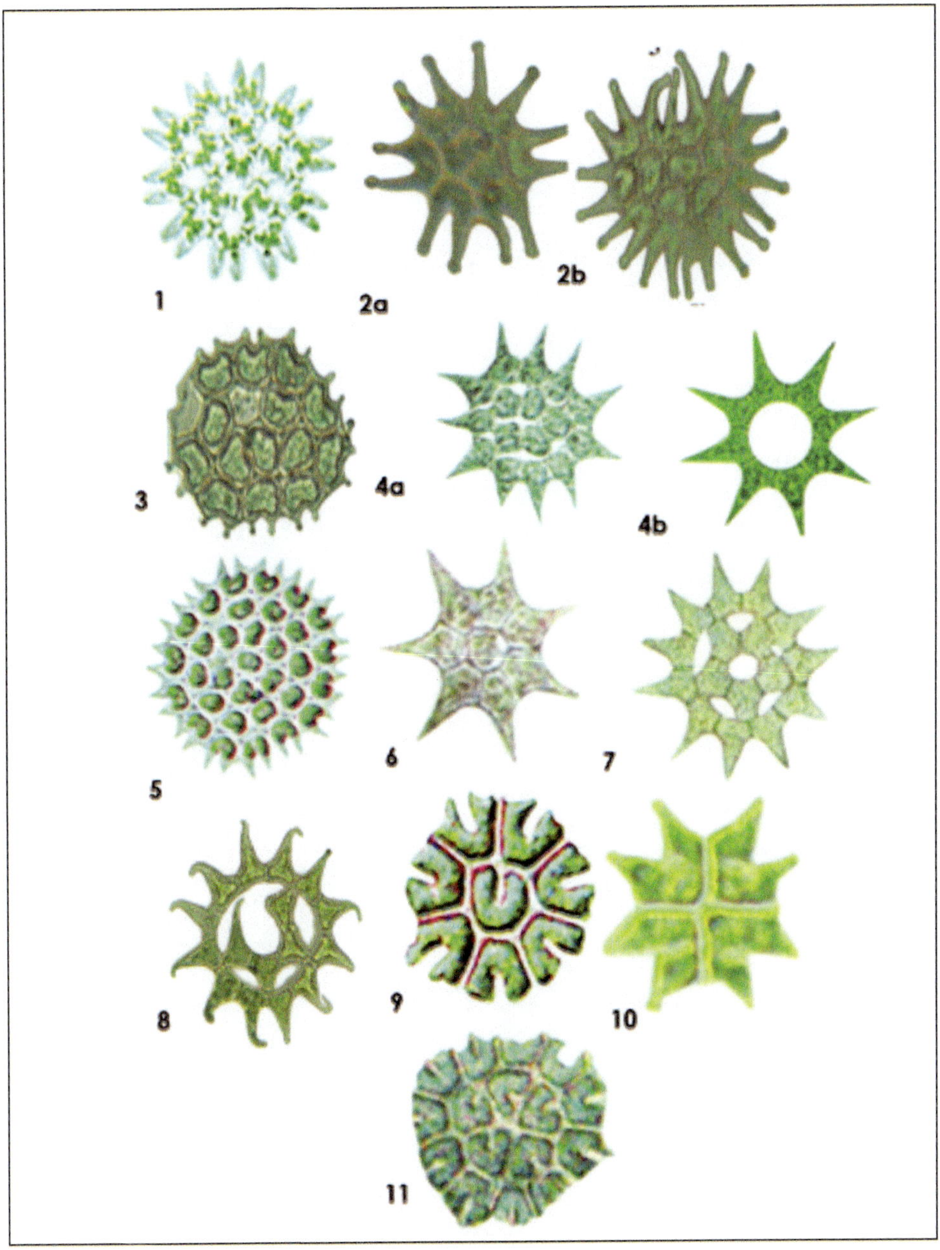


**Plate 3**

**Fig. (1) *P. duplex v. subgranulatum* (2) *P. glanduliferum* (3) *P. integrum* (4) *P. ovatum* (5) *P. sculptatum* (6) *P. simplex* (7) *P. simplex v. duodenarium* (8) *P. simplex* (9) *P. tetras* (10) *P. tetras v. excisum* (11) *P. tetras v. tetraodon***

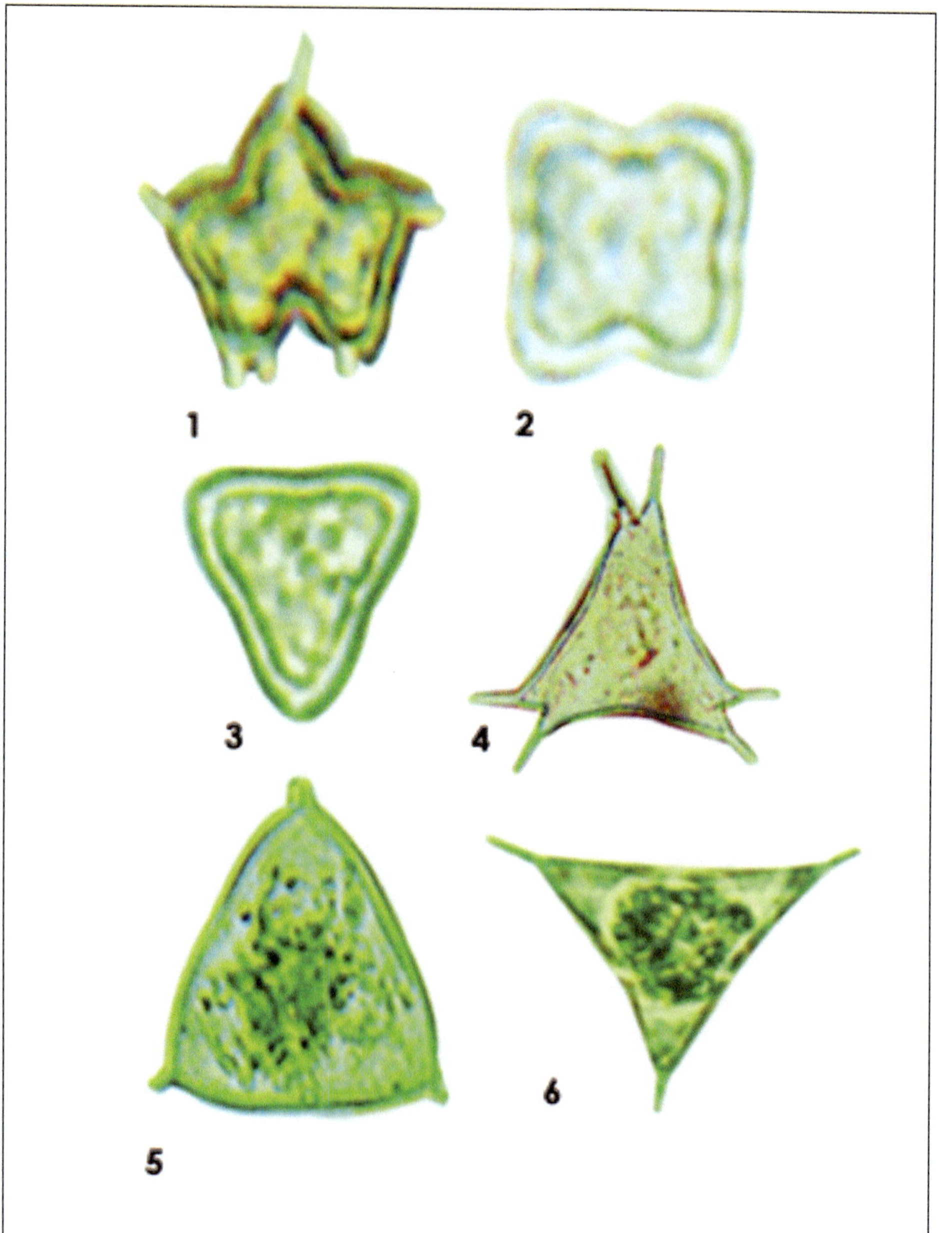


**Plate 4**

**Fig. (1) *Tetraedron caudatum* (2) *T. minimum* (3) *T. muticum f. punctulatum* (4) *T. regulare v. bifurcatum* (5) *T. regulare v. granulata* (6) *T. trigonum***

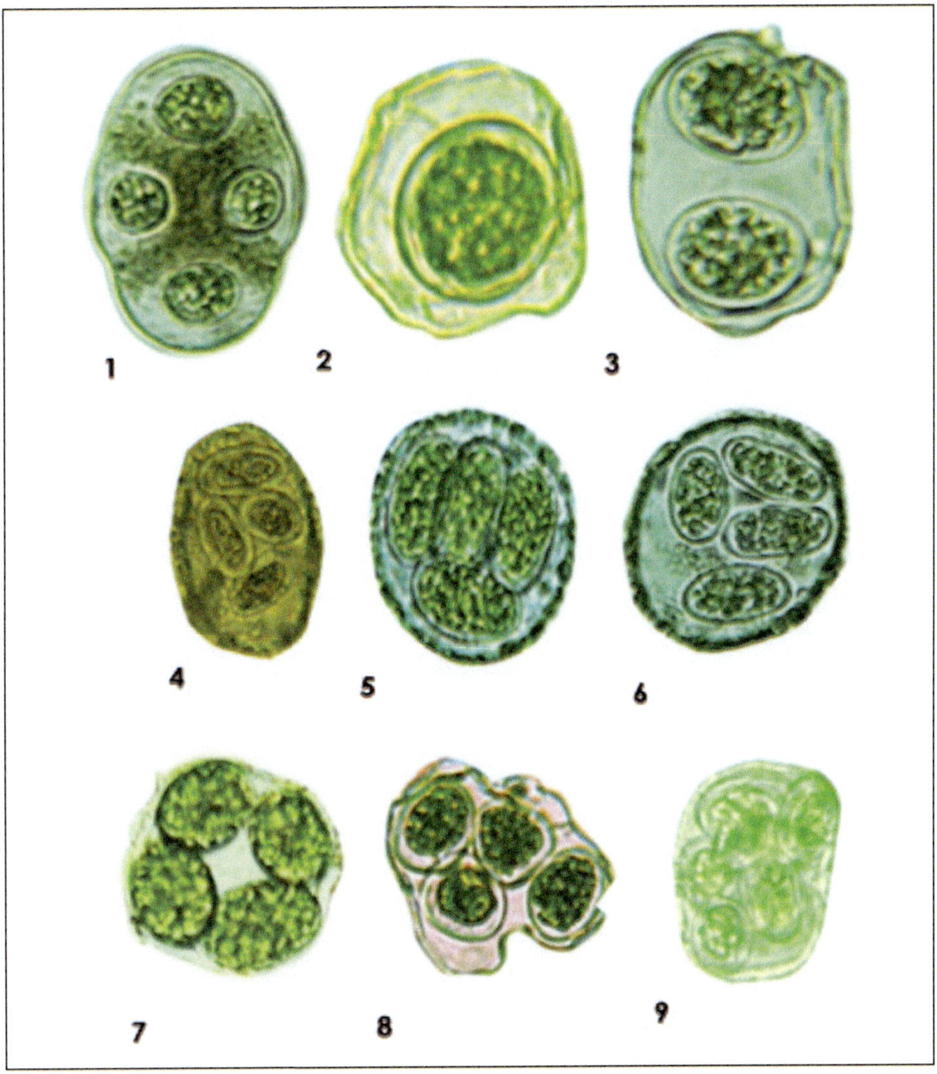


**Plate 5**

**Fig. (1) *Oocystis borgei* (2) *O. crassa* (3) *O. ecballocystiformis* (4) *O. elliptica* (5) *O. gigas* (6) *O. irregularis* (7) *O. naegelii A* (8) *O. pyriformis* (9) *O. solitaria***

**Plate 6**

**Fig. (1) *Selenastrum bibraianum* (2) *S. gracile* (3) *Ankistrodesmus falcatus* (4) *A. falcatus v. radiatus* (5) *A. spiralis v. fasciculatus* (6) *Kirchneriella contorta* (7) *K. lunaris* (8) *Actinastrum hantzschii***

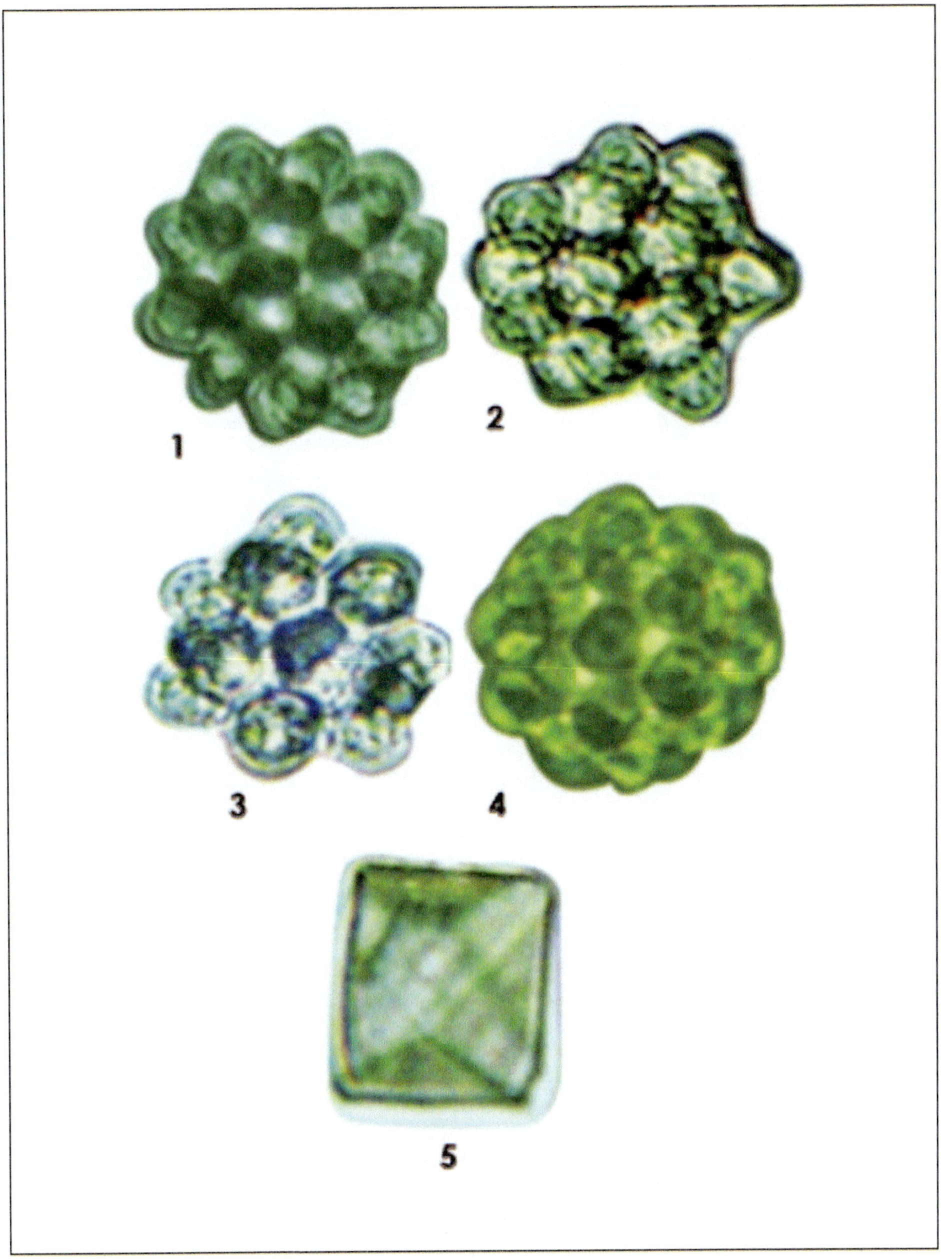


Plate 7

Fig. (1) *Coelastrum cambricum* v. *intermedium* (2) *C. microporum* (3) *Ñ. proboscideum* (4) *C. scabrum* (5) *Crucigeniate trapedia*

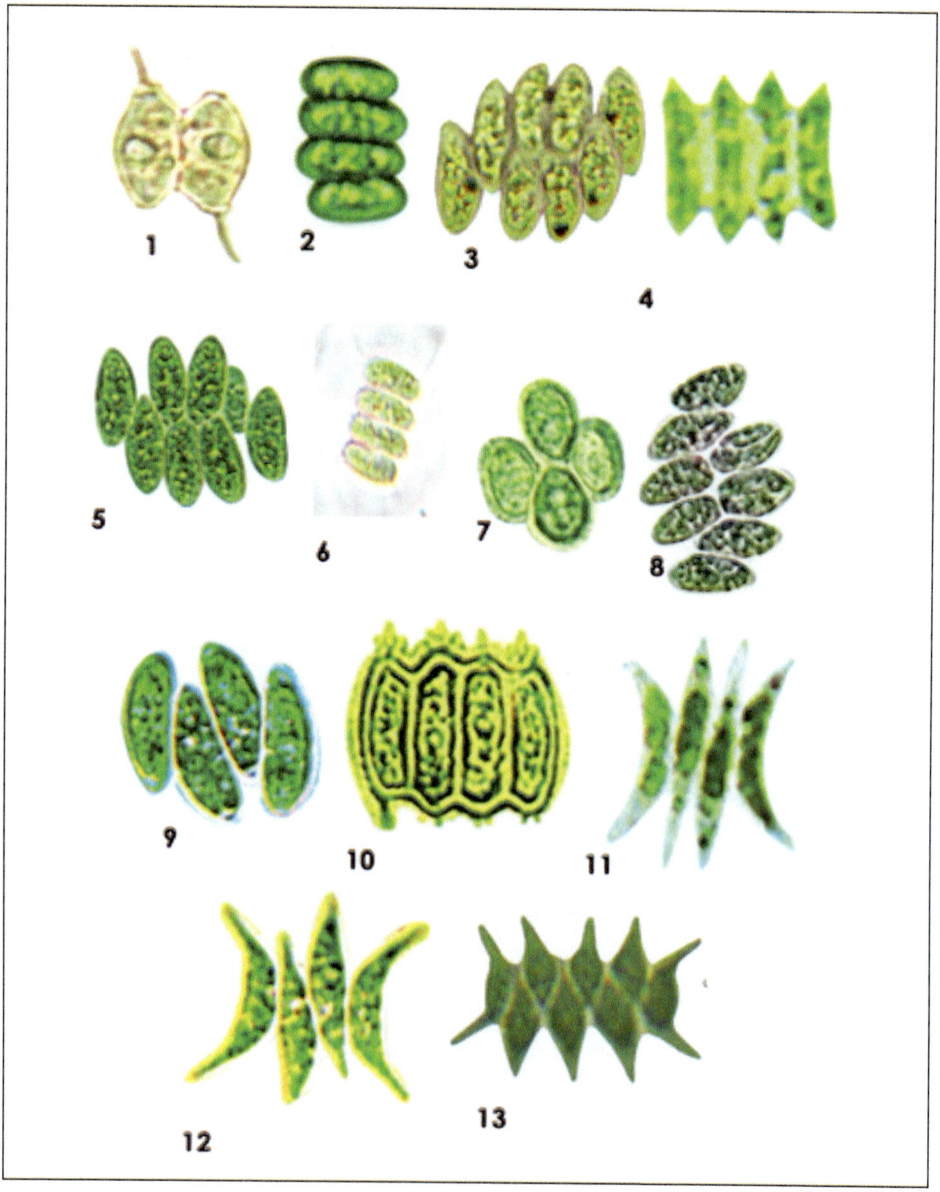


**Plate 8**

**Fig. (1) *Scenedesmus acutiformis* (2) *S. arcuatus* (3) *S. arcuatus v. capitatus* (4) *S. armatus v. major* (5) *S. armatus v. bicaudatus* (6) *S. bijuga* (7) *S. bijugatus v. flexuosus* (8) *S. bijugatus f. irregularis* (9) *S. bijugatus* (10) *S. brasiliensis* (11) *S. dimorphus* (12) *S. dimorphus f. tortus* (13) *S. falcatus v. major***

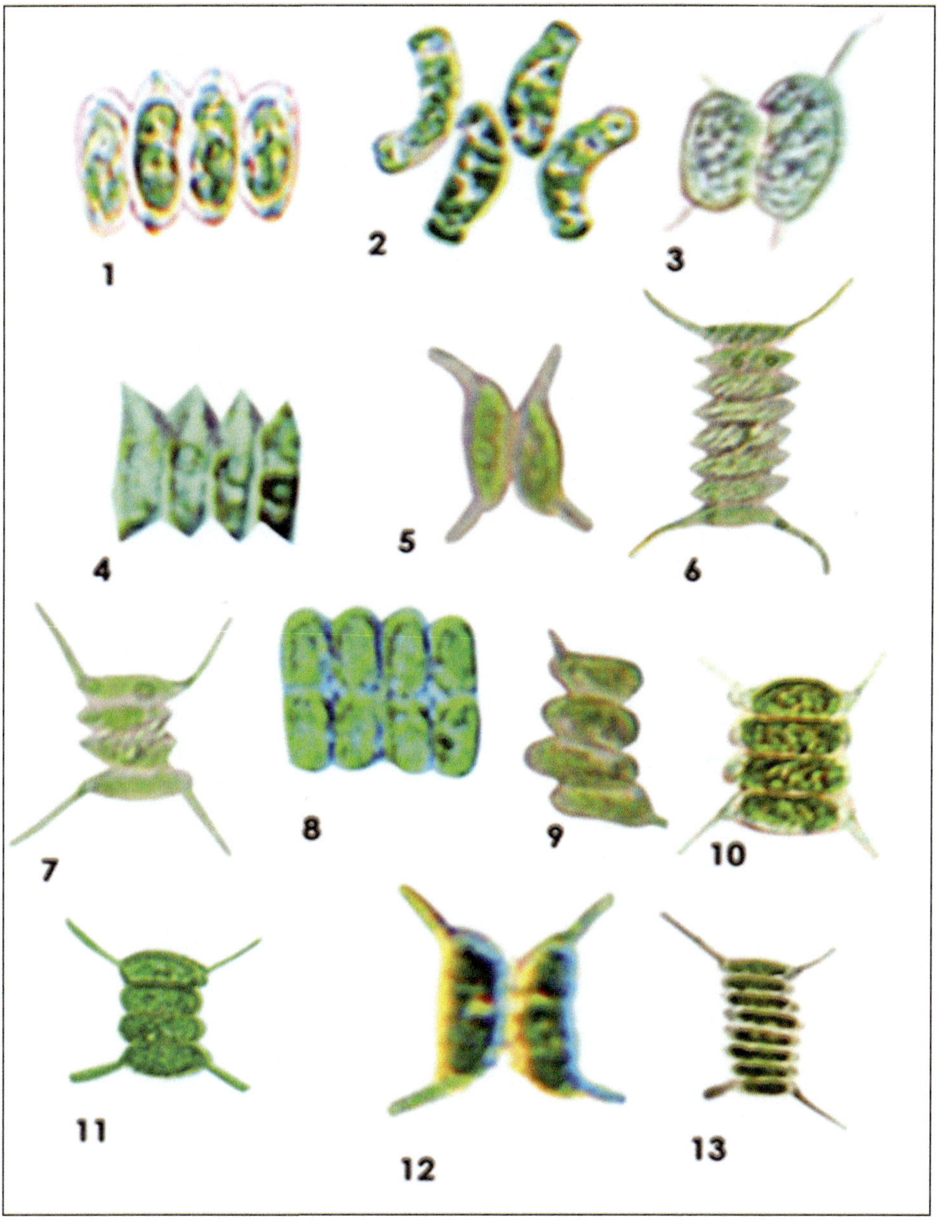

**Plate 9**

**Fig. (1) *S. incrassatulus v. mononae* (2) *S. indicus* (3) *S. longus v. naegelii* (4) *S. obliquus* (5) *S. opoliensis* (6) *S. opoliensis v. mononensis* (7) *S. protuberans f. minor* (8) *S. quadricauda v. bicaudatus* (9) *S. platydiscus* (10) *S. quadricauda v. longispina* (11) *S. quadricauda v. maximus* (12) *S. quadricauda v. quadrispina* (13) *S. quadricauda v. westii***

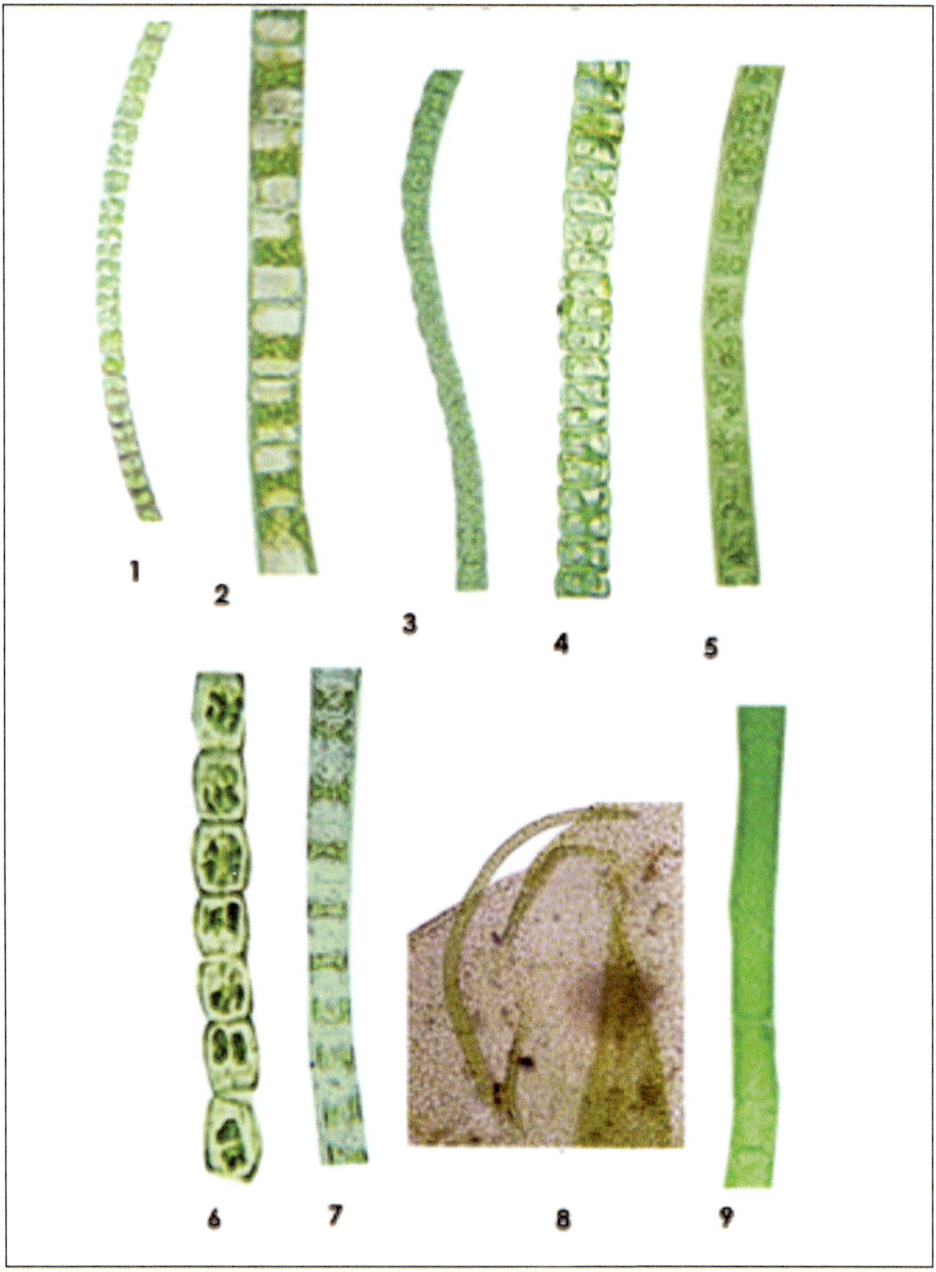

**Plate 10**

**Fig. (1) *Uronema confervicolum* (2) *Ulothrix subtilissima* (3) *U. tenerrima* (4) *U.tenuissima* (5) *U. variablis* (6) *U. zonata* (7) *Geminella minor* (8) *Schizomeris leibleinii* (9) *Rhizoclonium hieroglyphicum***

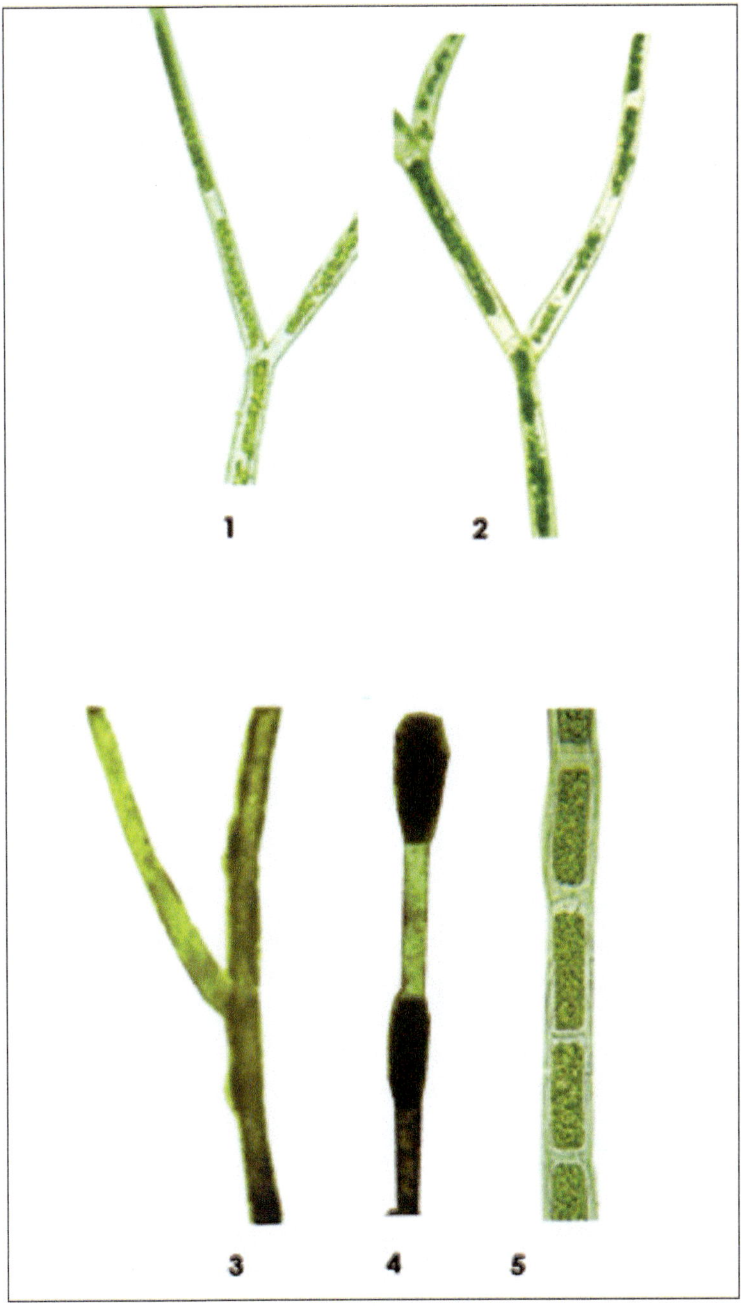

**Plate 11**

Fig. (1) *Cladophora fracta* (2) *C. fracta v. lacustris* (3) *C. insignis* (4) *Pithophora oedogonia* (5) *Basicladia chelonum*

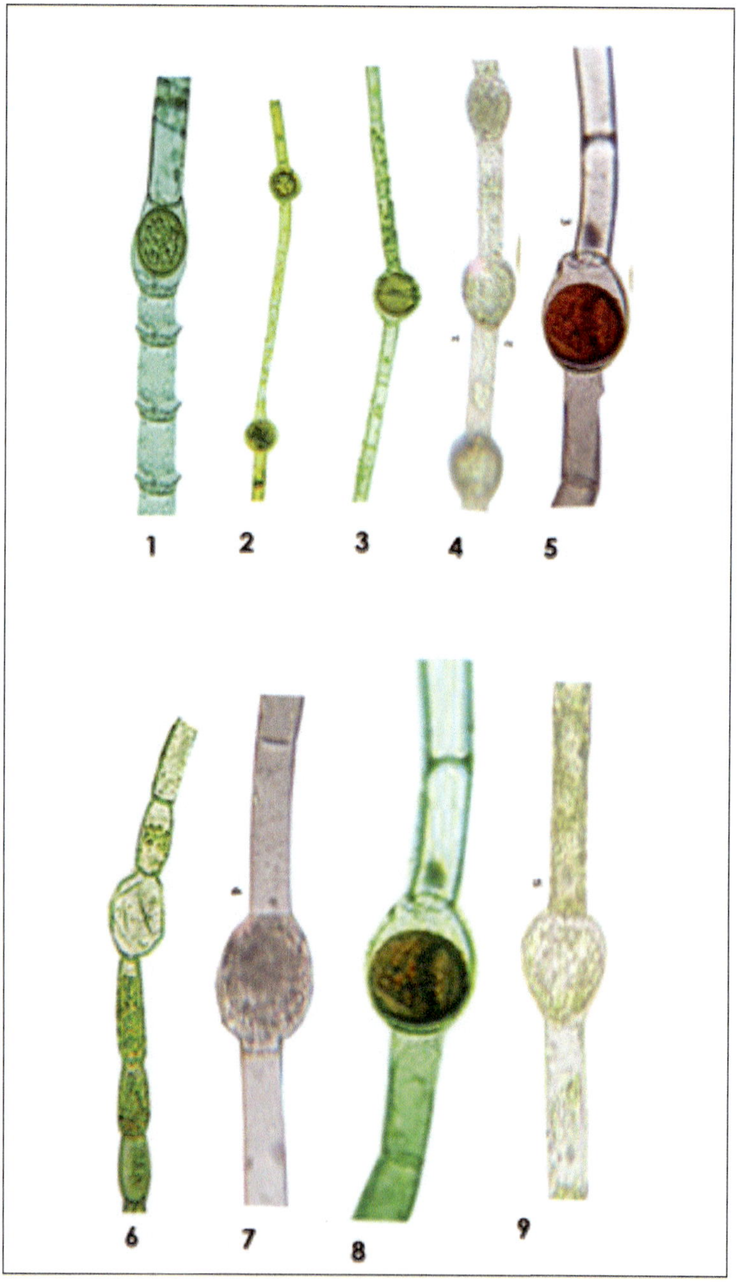


**Plate 12**

**Fig. (1) *Oedogonium capilliforme v. australe* (2) *O. cardiacum* (3) *O. cymatosporum* (4) *O. gallicum* (5) 0. *glabrum f. glabrum* (6) *O. globosum* (7) *O. kufferathii* (8) *O. patulum* (9) *O. pseudotumidulum***

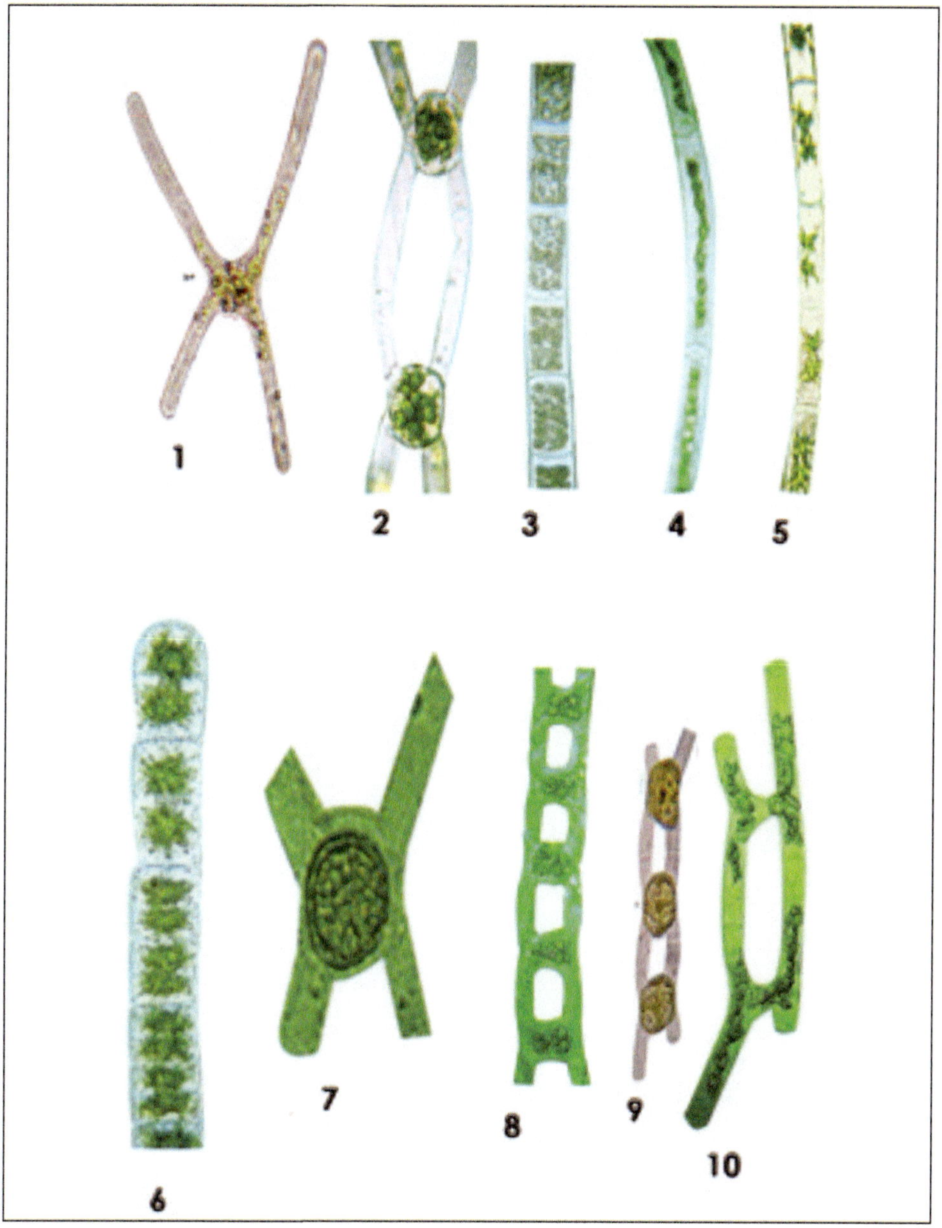


**Plate 13**

**Fig. (1) *Mougeotia drouetii* (2) *M. calcarea* (3) *M. scalaris* (4) *M.sphaerocarpa* (5) *Zygnemakhannae* (6) *Z. pectinatum* (7) *Z.synadelphum* (8) *Zygnemopsis jogensis* (9) *Zygnemopsis stephensiae* (10) *Debarya costata***

**Plate 14**

**Fig. (1) *Spirogyra aplanospora* (2) *Spirogyra borgeana* (3) *S. collinsii* (4) *S. crassa* v. *minispora* (5) *S.condensata* (6) *S. fallax* (7) *S. gracilis* (8) *S. gratiana* (9) *S. hyalina* (10) *S.irregularis***

**Plate 15**

**Fig. (1) *S. juergensii* (2) *S. neglecta* (3) *S. orientalis* (4) *S. parvispora* (5) *S. parvula* (6) *S. plena* (7) *S. pratensis* (8) *S. propria* (9) *S. pseudofloridana* (10) *S. pulchrifigurata***

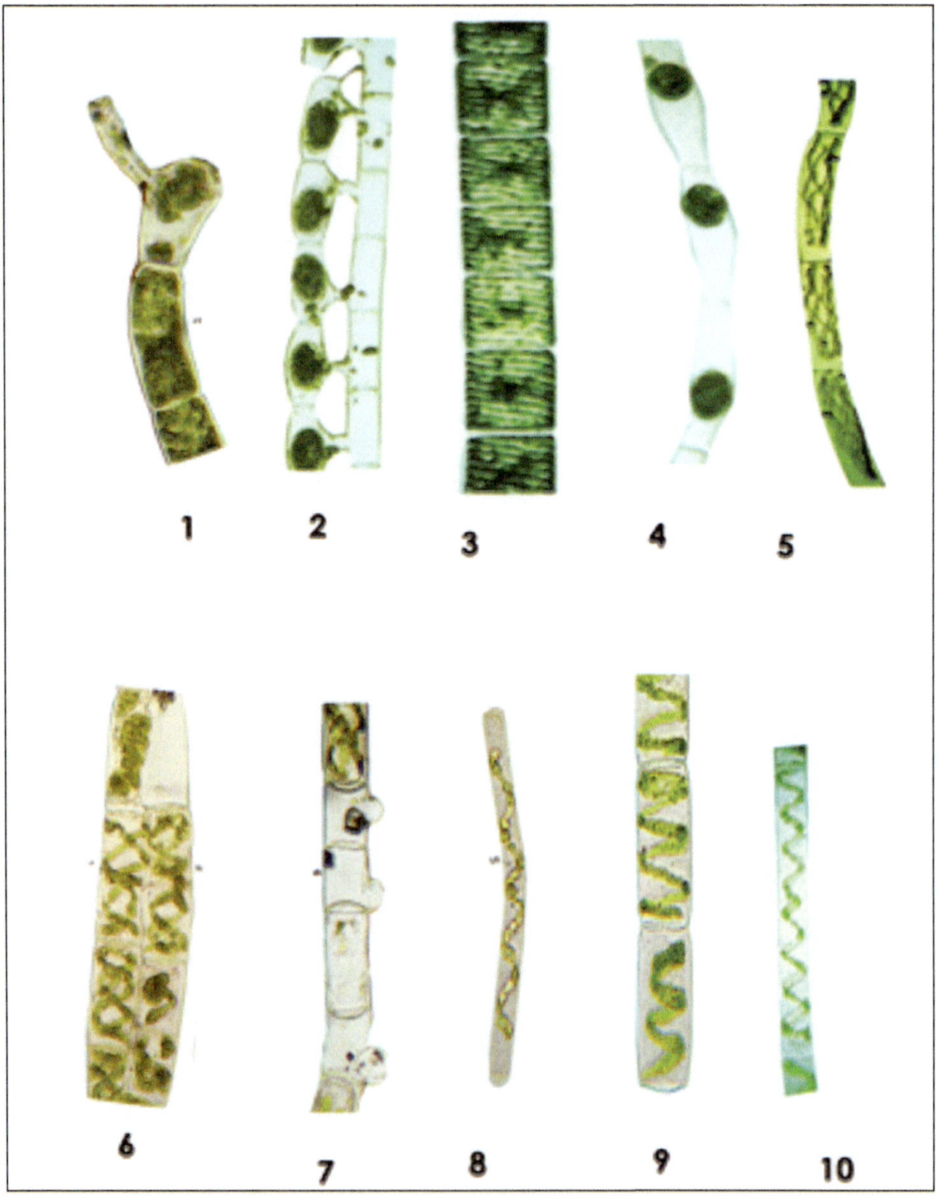

**Plate 16**

**Fig. (1) *S. rhizobrachialis* (2) *S. rivularis* (3) *S.setiformis* (4) *S. spreeiana* (5) *S. stictica* (6) *S. subsalsa* (7) *S. triplicata* (8) *S.undulisepta* (9) *S. variformis* (10) *S. weberi***

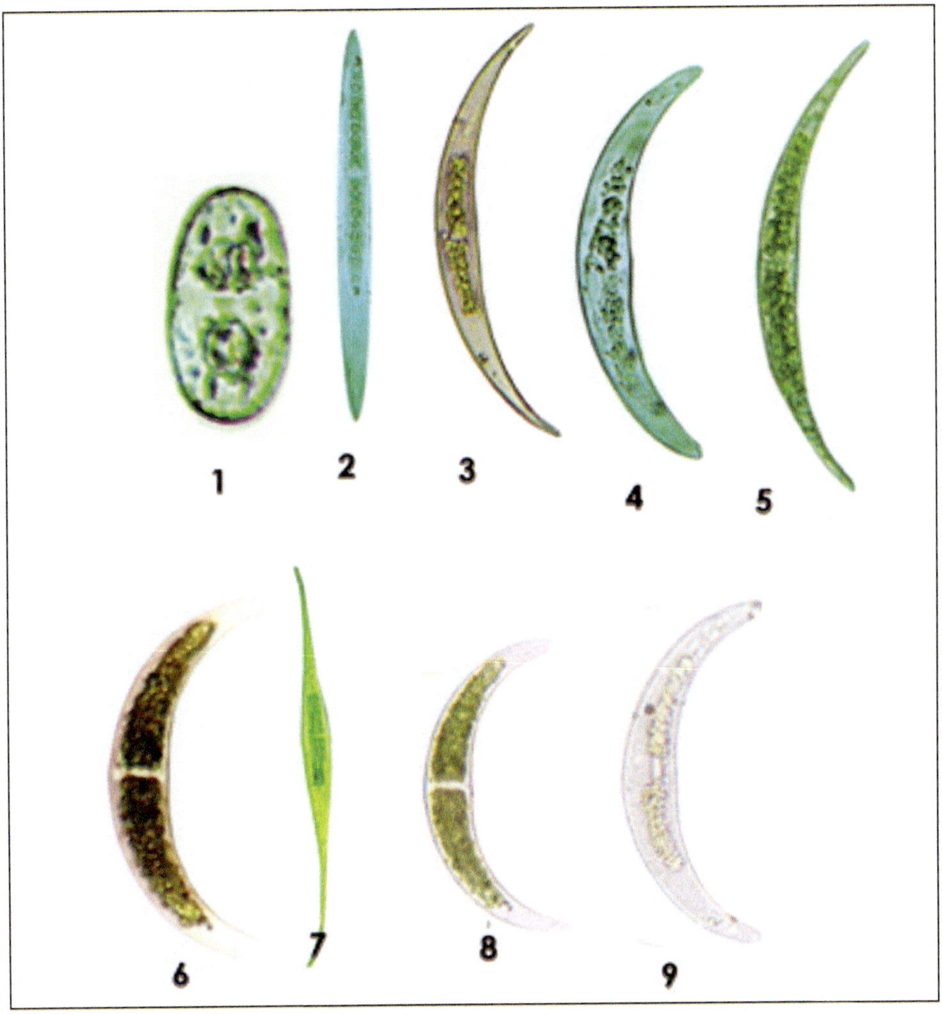


**Plate 17**

**Fig. (1)** ***Cylindrocystis subpyramidata*** **(2)** ***Closterium acerosum v. elongatum*** **(3)** ***C. acutum*** **(4)** ***C. calosporum v. maius*** **(5)** ***C. dianae*** **(6)** ***C. incurvum*** **(7)** ***C.kuetzingii*** **(8)** ***C. leibleinii*** **(9)** ***C. libellula intermedium***

**Plate 18**

**Fig. (1)** ***C. moniliferum*** **(2)** ***C. parvulum*** **(3)** ***C. parvulum v. angustum*** **(4)** ***Ñ.peracerosum*** **(5)** ***C. rectimarginatum*** **(6)** ***C. rectimarginatum v. andamense*** **(7)** ***C. sigmoideum*** **(8)** ***C. cynthia***

**Plate 19**

**Fig. (1) *Penium spinospermum* (2) *Pleurotaenium ehrenbergii* (3) *Micrasterias foliacea* (4) *M. mahabuleshwarensis* (5) *M. pinnatifida* (6) *M. radians* (7) *M. zeylanica***

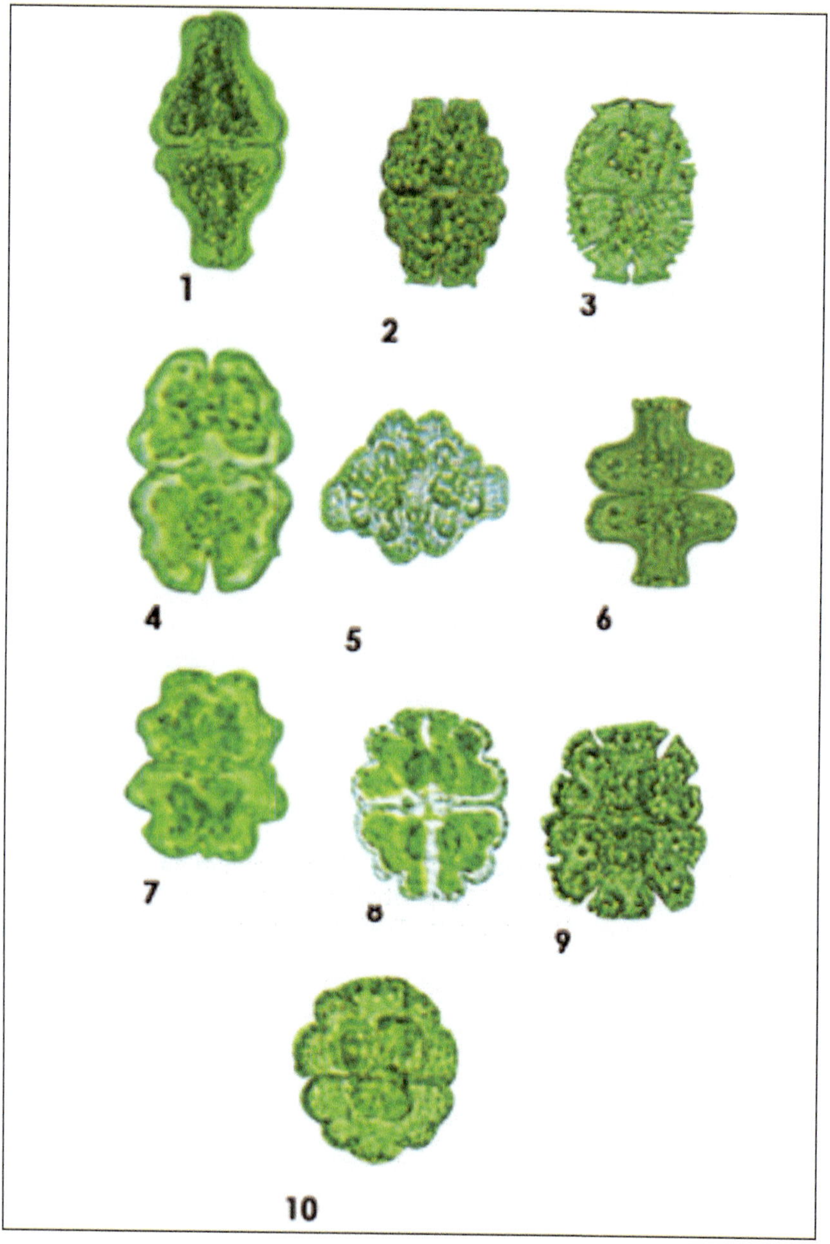

**Plate 20**

**Fig. (1) *Euastrum mansatum v. pyxidatum* (2) *E. bidentatum* (3) *Å.coralloides v. trigibberum* (4) *E. elegans* (5) *E. irregulare* (6) *E.platycerum* (7)*E.pseudotuddalense* (8) *E. spinulosum* (9) *E.spinulosum v. inermius* (10) *E. spinulosum v. burmense***

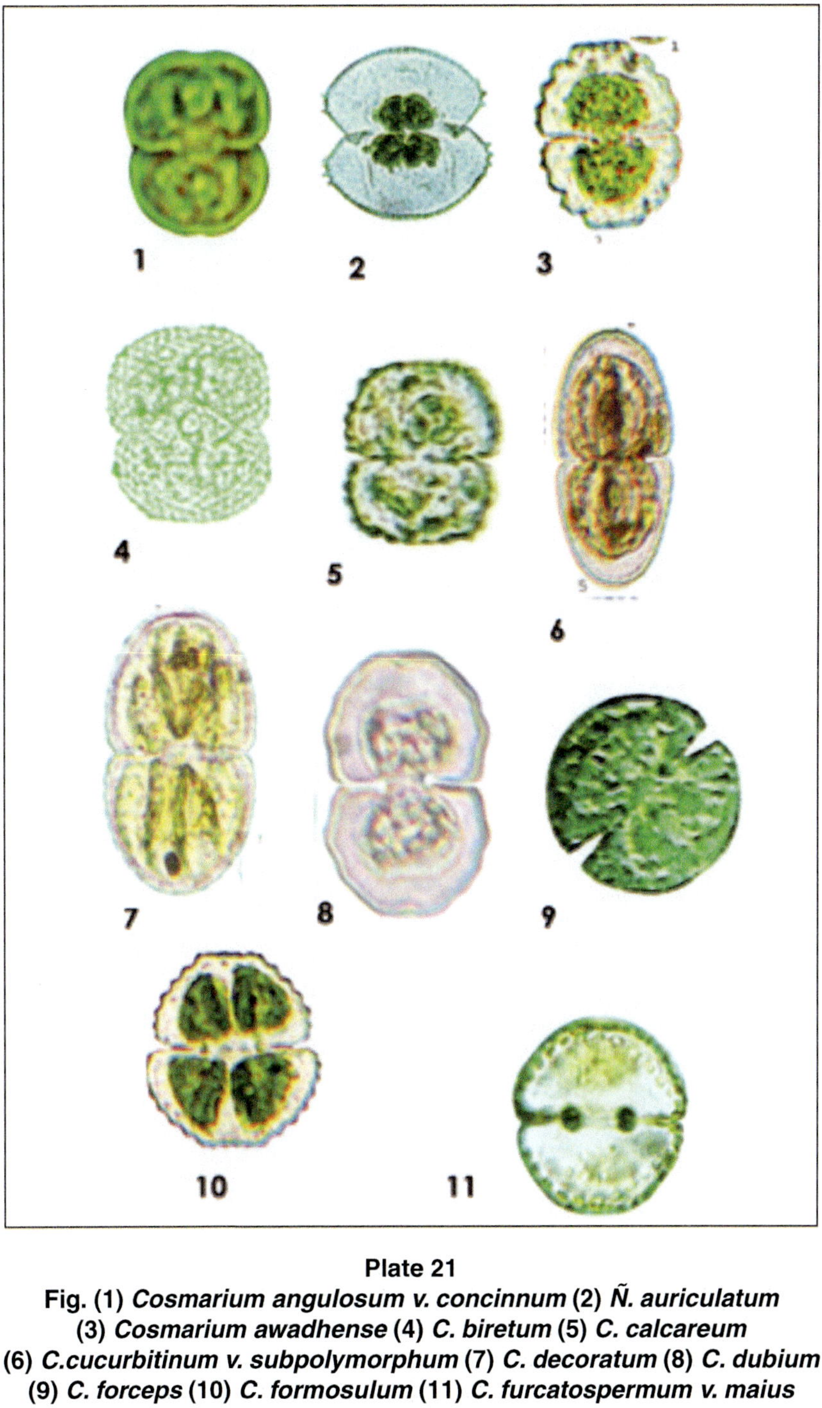


**Plate 21**

**Fig. (1) *Cosmarium angulosum v. concinnum* (2) *Ñ. auriculatum* (3) *Cosmarium awadhense* (4) *C. biretum* (5) *C. calcareum* (6) *C.cucurbitinum v. subpolymorphum* (7) *C. decoratum* (8) *C. dubium* (9) *C. forceps* (10) *C. formosulum* (11) *C. furcatospermum v. maius***

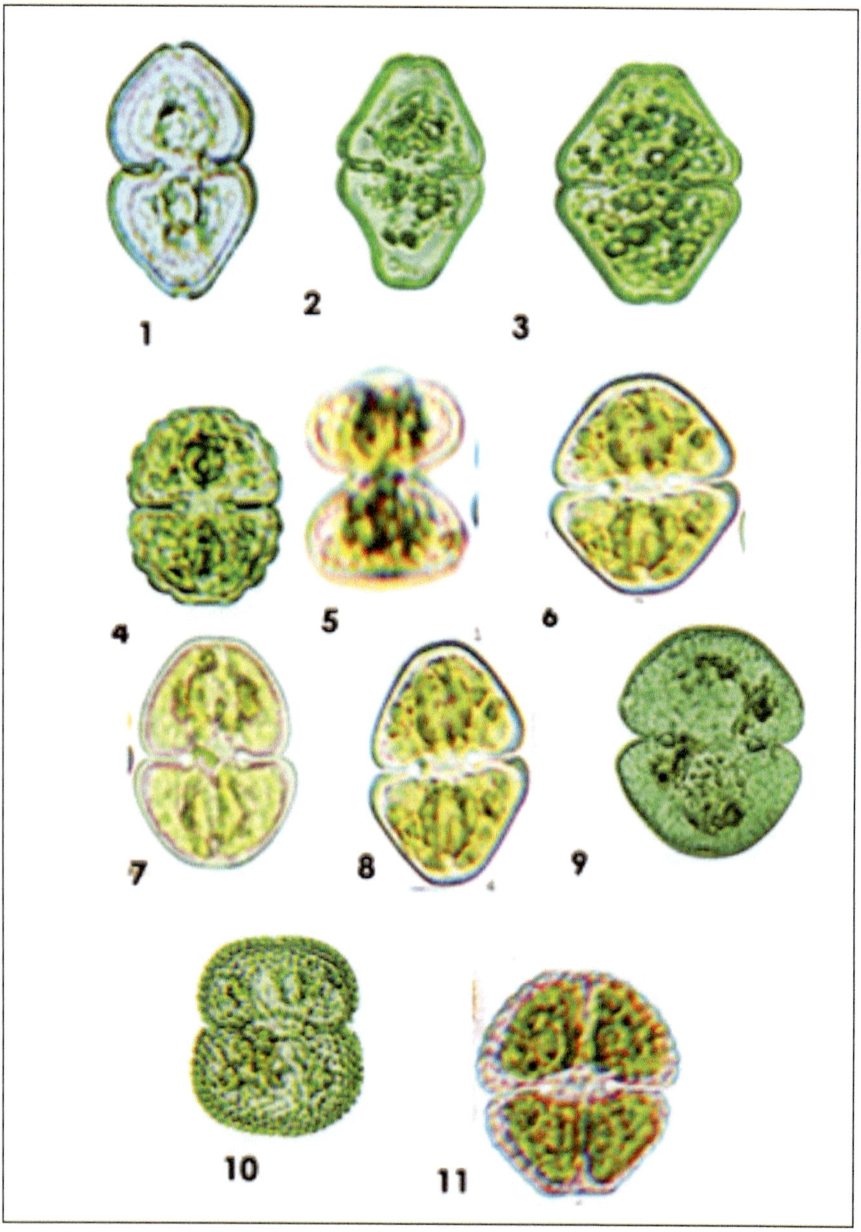


**Plate 22**

Fig. (1) *Cosmarium granatum* (2) *C. hammeri v. homalodermuò* (3) *Ñ. hammeri v. schmidlei* (4) *C. impressulum v. crenulatum* (5) *C. indentatum v. ellipticum* (6) *C. lundellii* (7) *C. lundellii v. corruptum* (8) *C. lundellii v. ellipticum* (9) *C. lundellii v. subellipticum* (10) *C. margaritatum* (11) *C. miscellum*

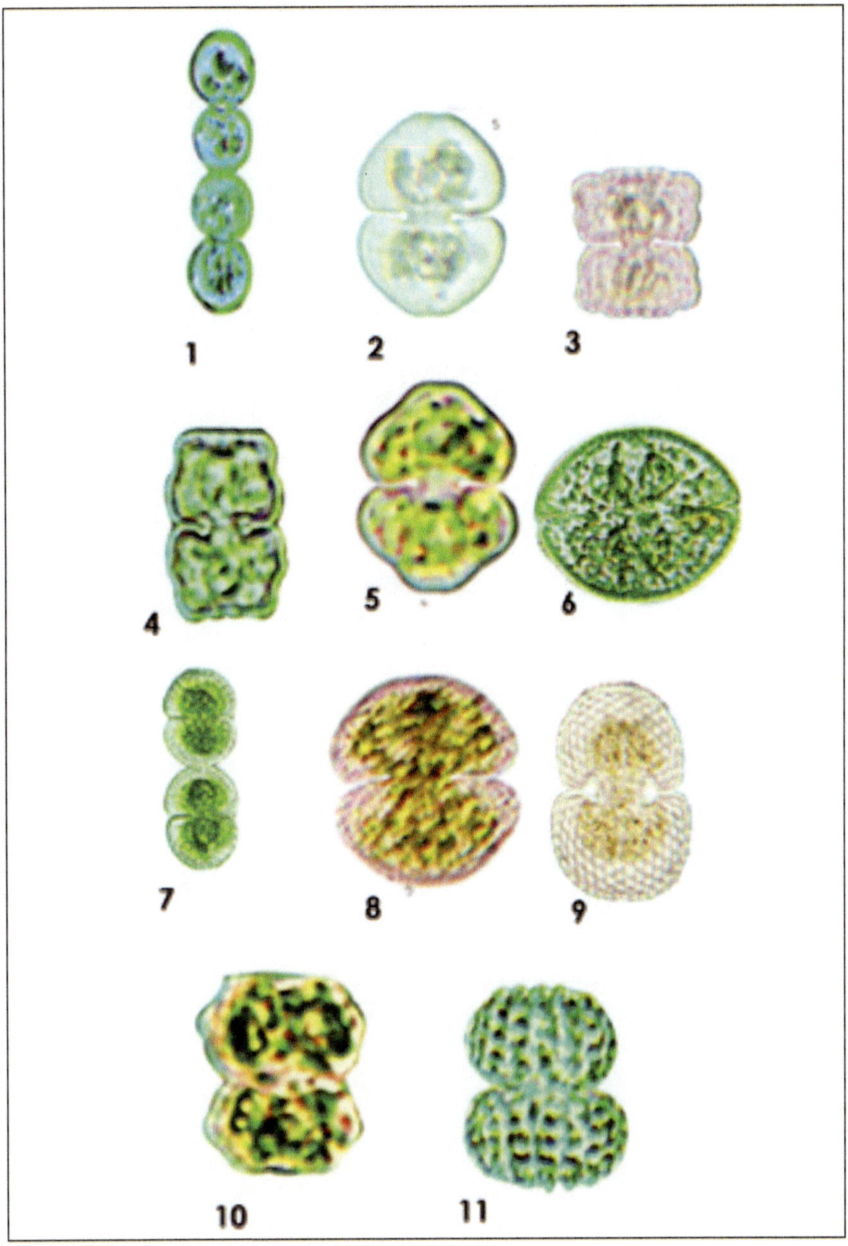


**Plate 23**

**Fig. (1) *Cosmarium moniliforme v. lamneticum* (2) *C. nitidulum* (3) *C. norimbergense v. depressum* (4) *C. norimbergense v. elongatum* (5) *C. nymannianum* (6) *Ñ. obsoletum v. sitvense* (7) *C. obtusatum* (8) *C. pachydermum v. aethiopicum* (9) *C.pardalis* (10) *C. polygonum* (11) *C. portianum v. nephroideum***

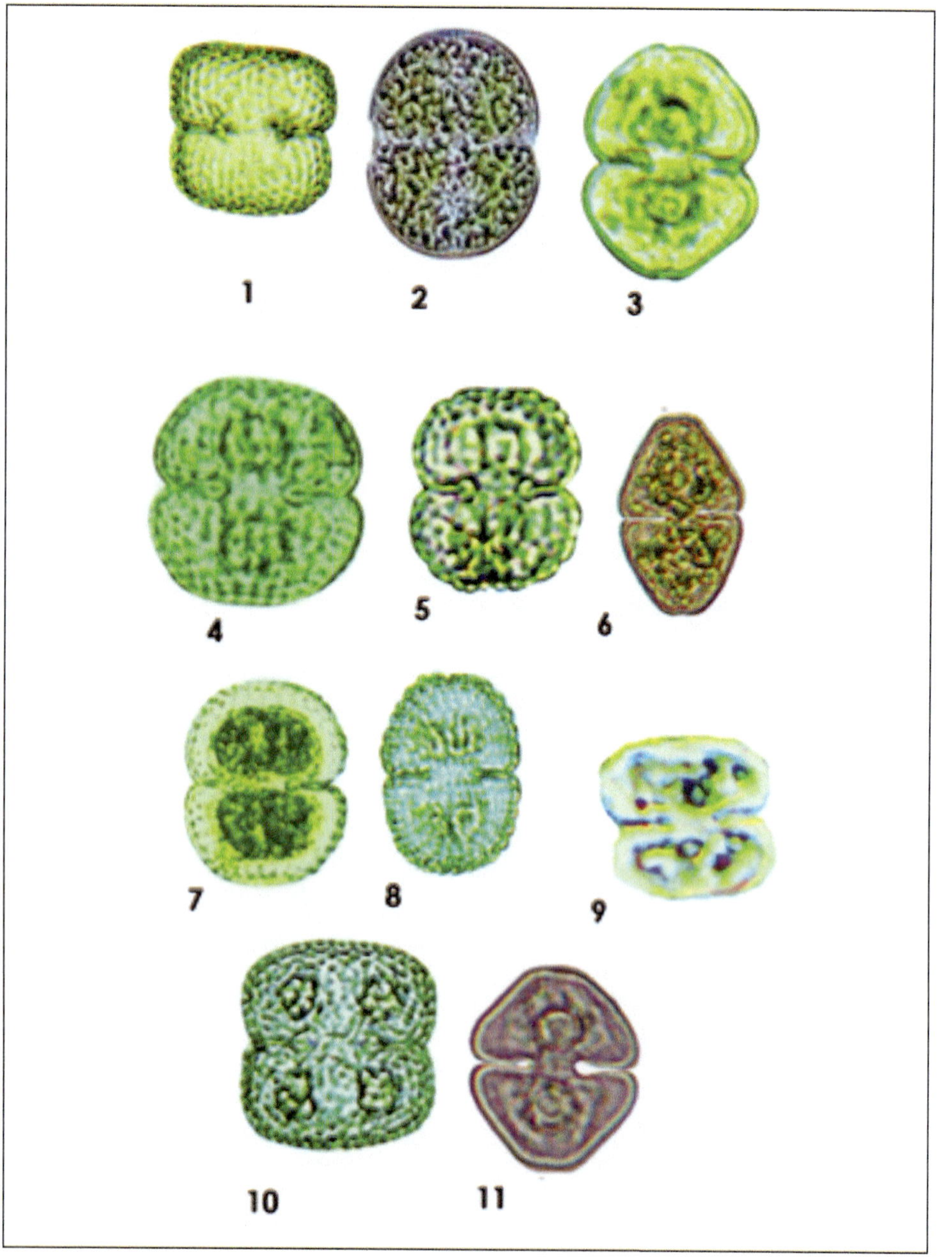


**Plate 24**

**Fig. (1)** ***Cosmarium pseudobroomei*** **(2)** ***C. pseudoconnatum*** **(3)** ***C.pseudogranatum v. rotundatum*** **(4)** ***C. punctulatum*** **(5)** ***C. punctulatum v. subpunctulatum*** **(6)** ***C. Pyramidatum*** **(7)** ***C. quinarium*** **(8)** ***C. radiosum*** **(9)** ***C. regnellii*** **(10)** ***C. reniforme*** **(11)** ***C. retusiforme***

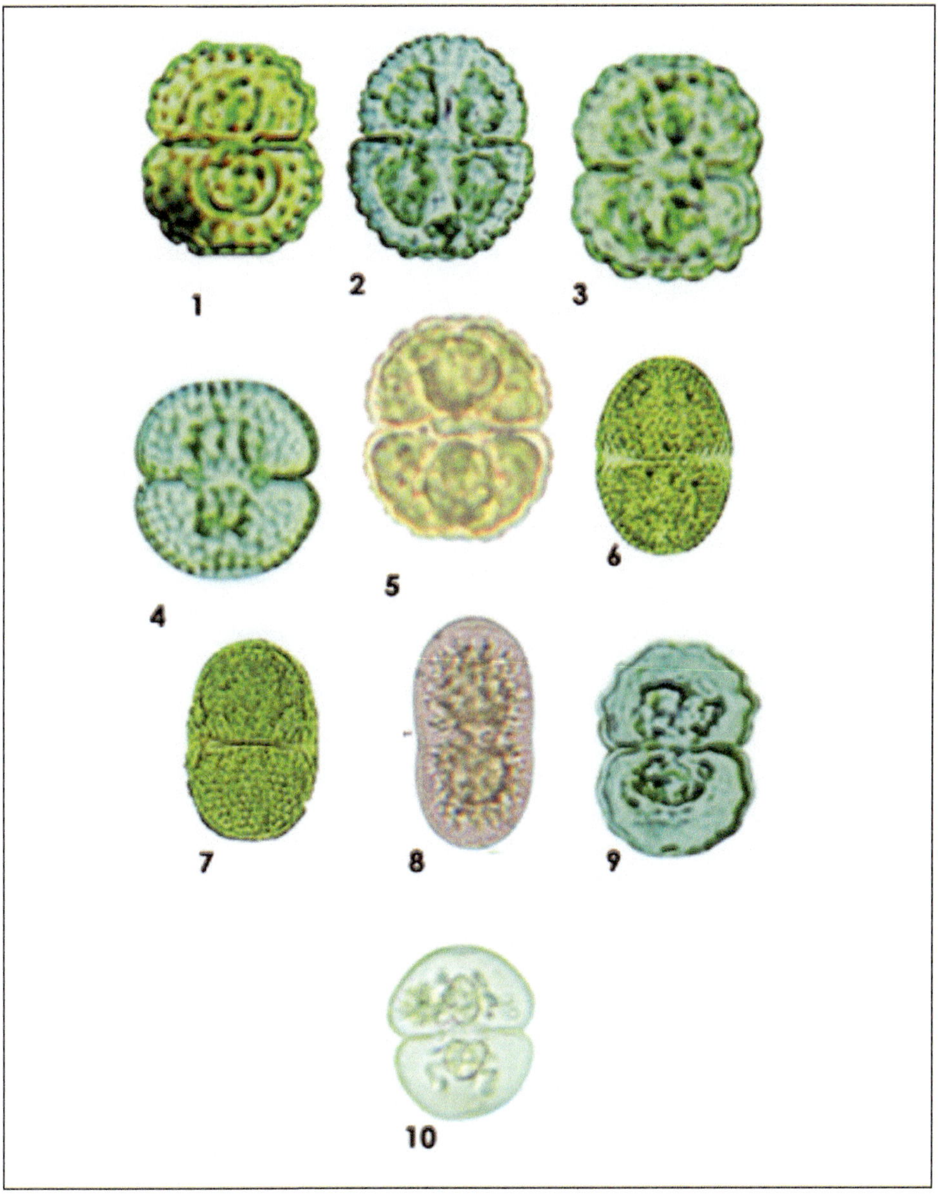

**Plate 25**

**Fig. (1) *C. sexnotatum* (2) *C. speciosum* (3) *C. subalatum* (4) *C. subcrenatum* (5) *C. subimpressulum* (6) *Ñ.turgidum v. minus* (7) *C. turgidum* v. *subrotundatum* (8) *C. turgidum v. westii* (9) *Ñ.undulatum* v. *minutum* (10) *C. variolatum v. rotundatum***

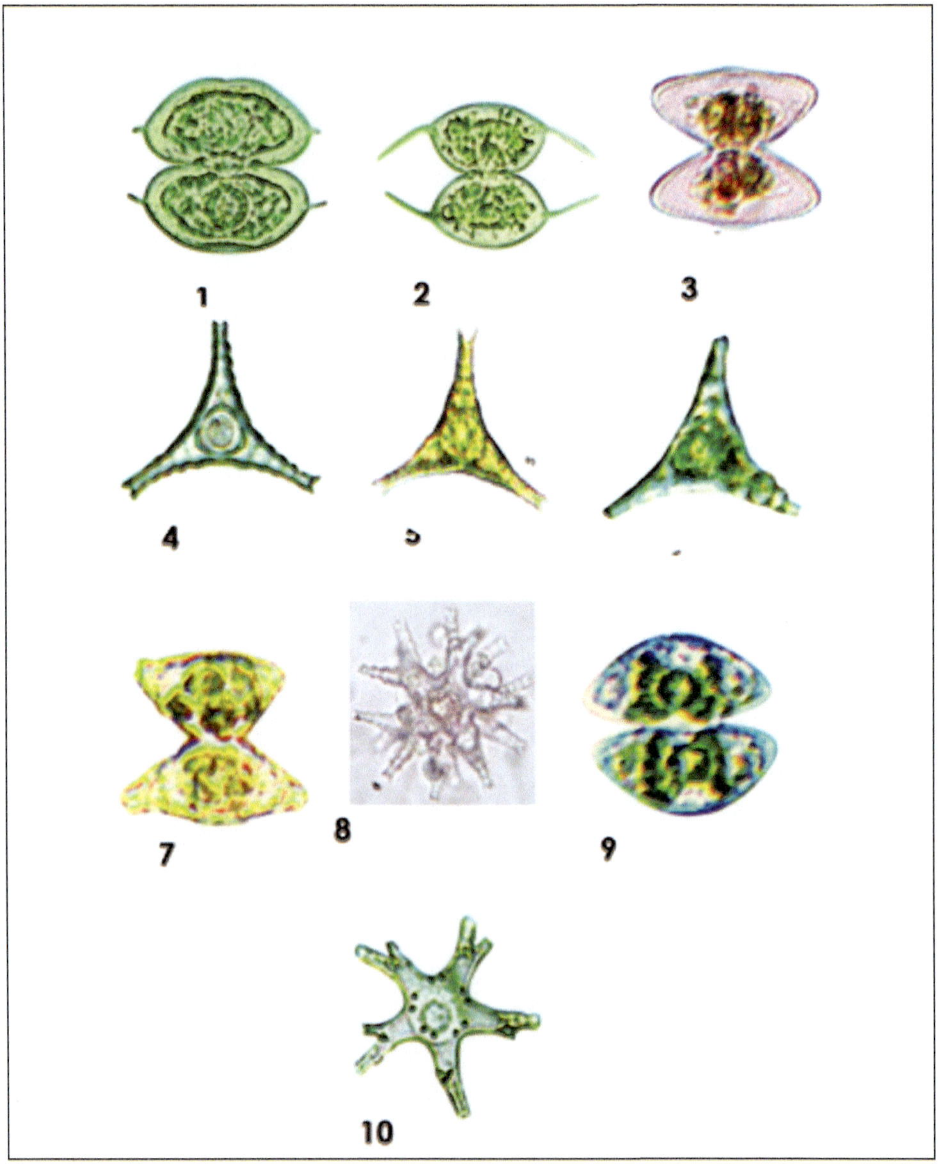

**Plate 26**

**Fig. (1) *Arthrodesmus convergens* (2) *A. curvatus f. major* (3)*Staurastrum bieneanum v. ellipticum* (4) *S. gracile* (5) *S. gracile v. coronulatum* (6) *S. gracile v. kriegeri* (7) *S. granulosum* (8) *S. leave* (9) *S.pachyrhynchum* (10) *S. sexangulare v. productum***

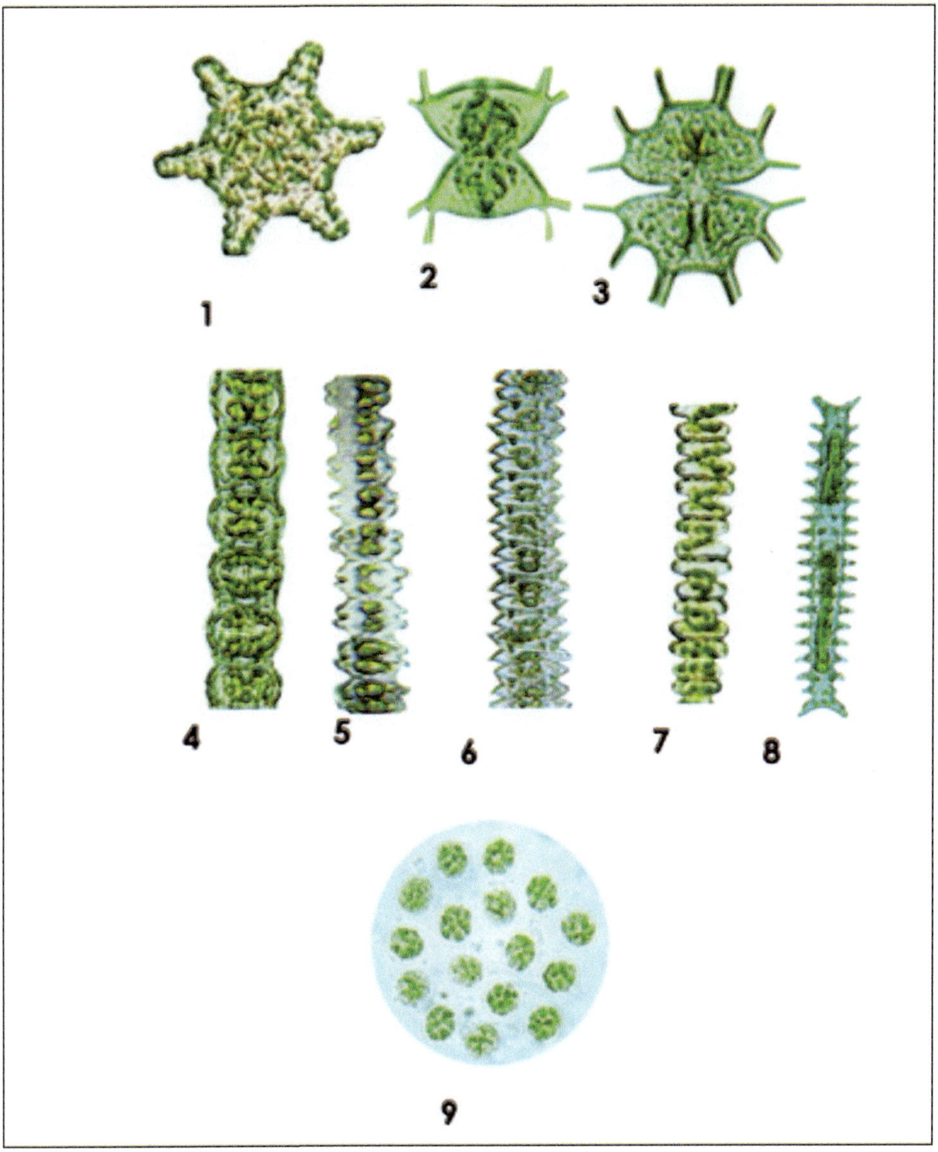


Plate 27

Fig. (1) *S. sexcostatum v. productum* (2) *S. wildemanii* (3) *Xanthidium sexmamillatum* (4) *Desmidiumbaileyi v. baileyi f. tetragonum* (5) *D. swartzii* (6) *Onychonema leave v. latum* (7) *Sphaerozosma excavata* (8) *Triploceras gracile* (9) *Asterococcus limneticus*

**Plate 28**

**Fig. (1) *Nitella bengalensis* (2) *N.microcarpa* (3) *Chara brachypus* (4) *C. braunii* (5) *C. delicatula* (6) *C. flaccida* (7) *C. hatei* (8) *C hydropitysf. major***

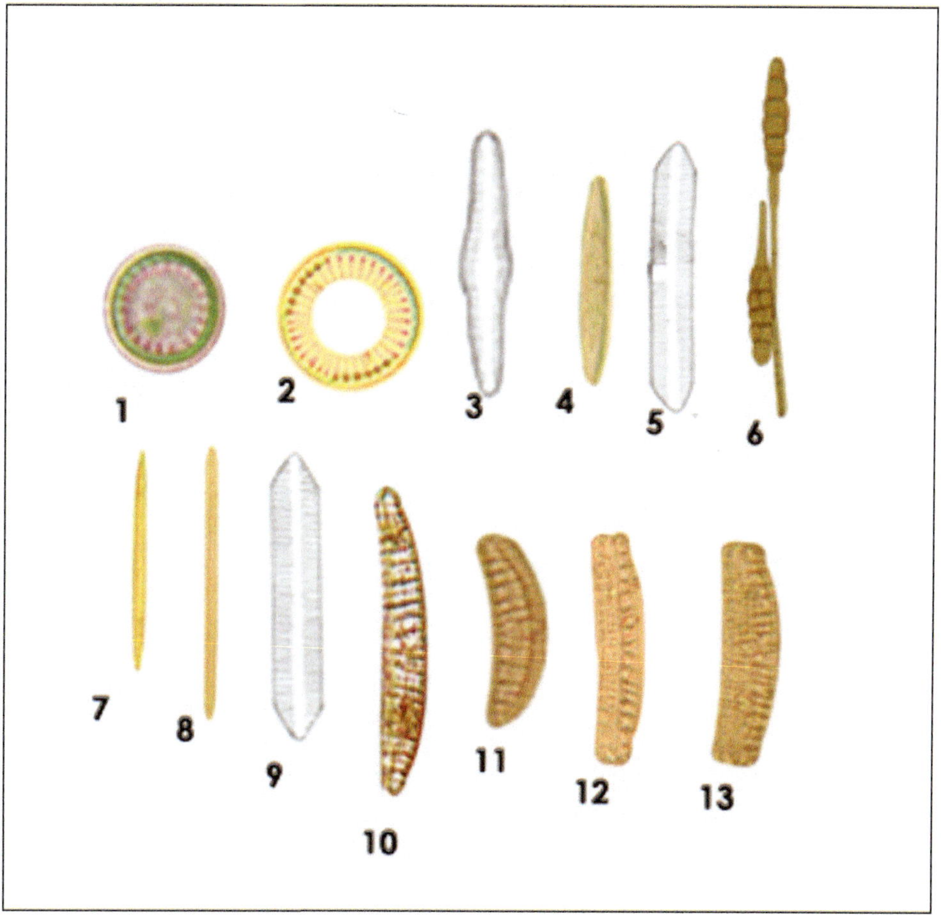


**Plate 29**

**Fig. (1)** ***Cyclotella glomerata*** **(2)** ***C. meneghiniana*** **(3)** ***Fragilaria brevistriata v. vidarbhensis*** **(4)** ***F. intermedia*** **(5)** ***F. intermedia f. chandrapurensis*** **(6)** ***Lichmophora abbreviata*** **(7)** ***Synedra acus v. acula*** **(8)** ***S. ulna*** **(9)** ***S. ulna v. notata*** **(10)** ***Eunotia arcus v. uncinata*** **(11)** ***E. lunaris v. subarcuata*** **(12)** ***E. major v.indica*** **(13)** ***E. praerupta***

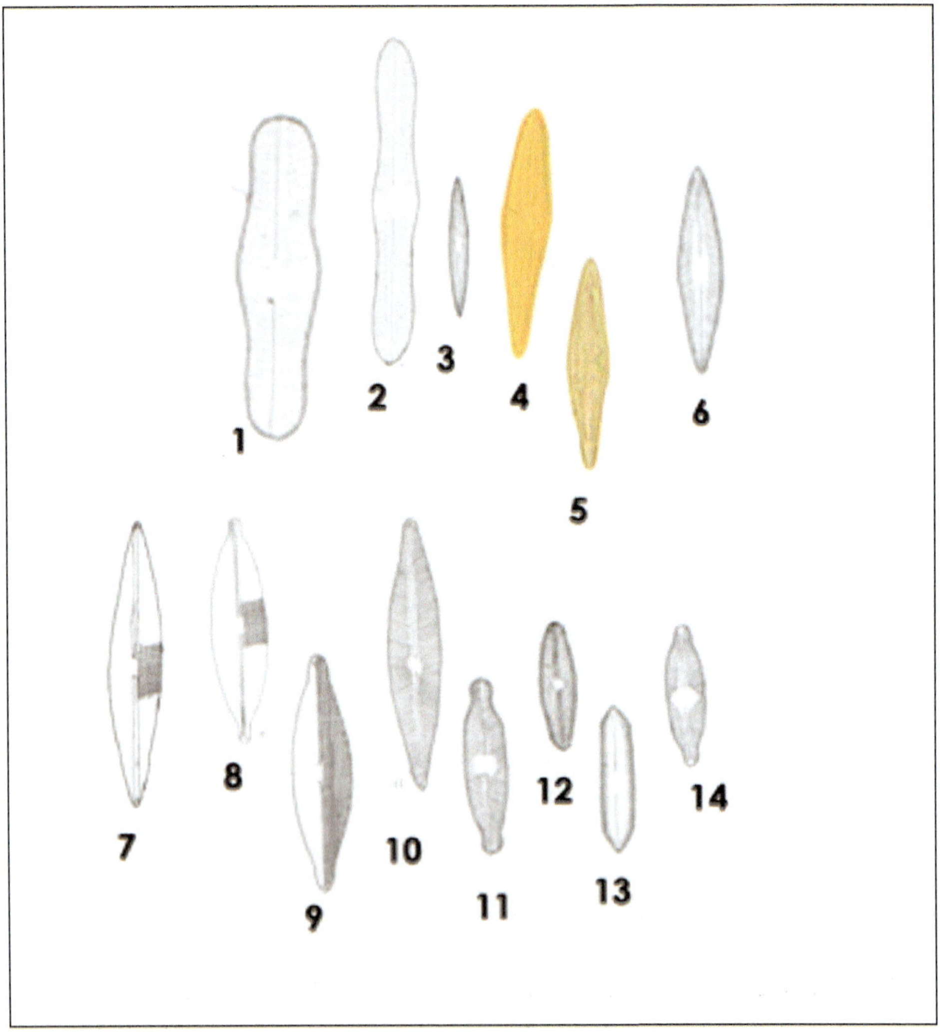


***Plate 30**

**Fig. (1) *Achnanthes inflata* (2) *A. inflata f. vidarbhensis* (3) *Navicula cari* (4) *N.cari v. angusta* (5) *N. carif. indica* (6) *N. cryptocephala* (7) *N. cuspidata* (8) *N. cuspidata v. ambigua f. diminuta.*9) *N. cuspidataf. brevirostrata* (10) *N. cuspidata v. heribaudi* (11) *N. dicephala v. sphaerophora* (12) *N. feverborni* (13) *N. iniqua* (14) *N. laterostrata***

***(*Source*: Andhale 2008).**

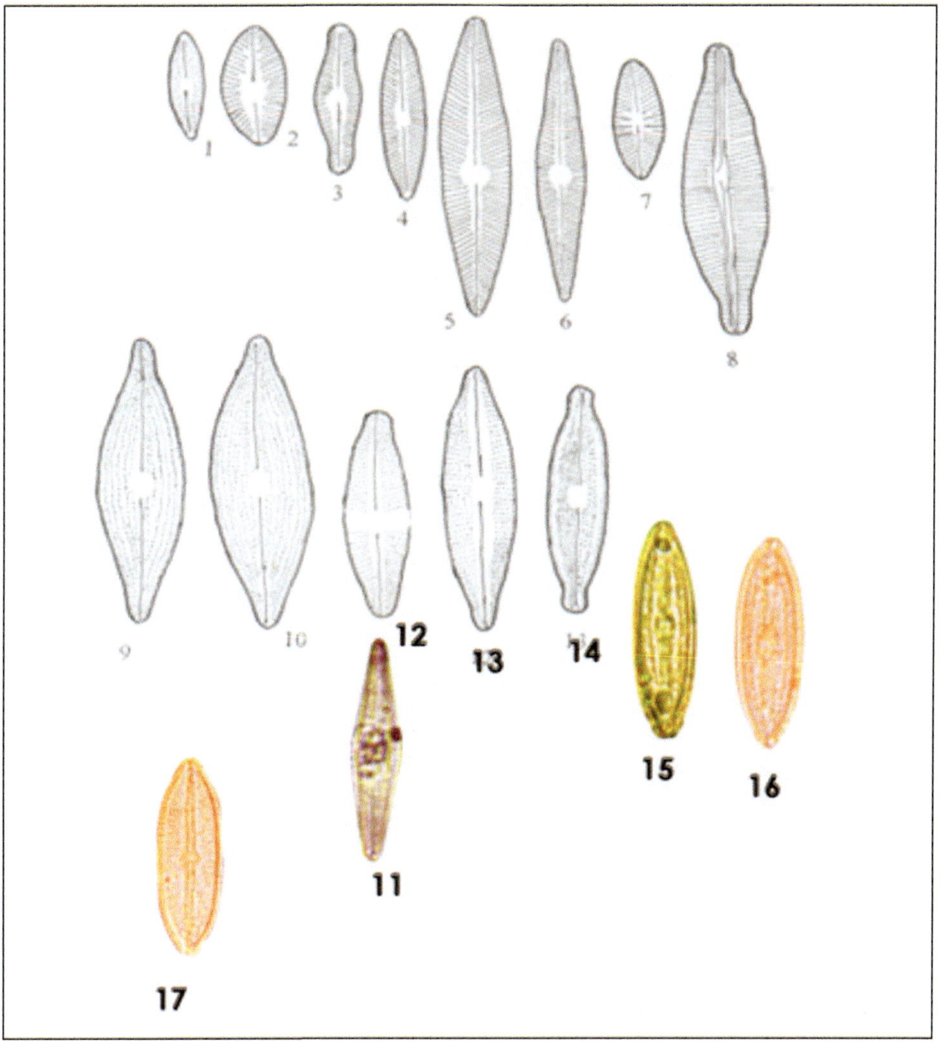


***Plate 31**

**Fig. (1) *N. mutica* (2) *N. pupula v. elliptica* (3) *N. pupula v. rostrata* (4) *N. radiosav.tenella* (5) *N. reinhardtii f. gracilior* (6) *N. rhynchocephala v. elongata* (7) *N. schonfeldii* (8) *N. varidula v. capitata* (9) *Anomoeoneis lanceolata* (10) *A.sculpta*11) *Stauroneis anceps v. hyalina* (12) *S. kirtikarii* (13) *S. phoenicenteron f. producta* (14) *Neidium affine v. amphirhynchus f. truncatula* (15) *Neidium amphigomphous v. obtusum* (16) *N. iridis f. ambigua* (17) *N. iridis f. dhulensis** (*Source*: Andhale 2008).**

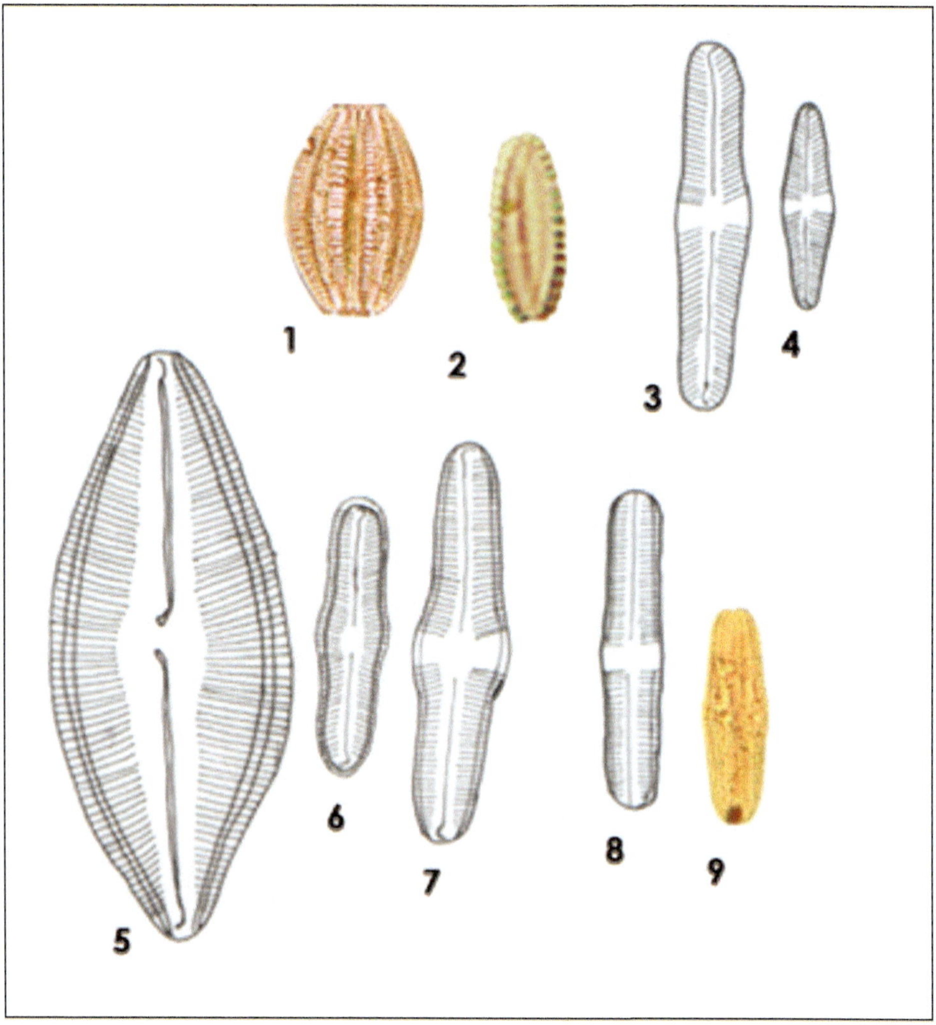


***Plate 32**

**Fig. (1) *Diploneis elliptica v. ladogensis* (2) *D. puella* (3) *Caloneis beccariana* (4) *C. beccariana v. hustedtii* (5) *C. permagna* (6) *C. silicula* (7) *C. silicula v. intermedia* (8) *C. silicula v. tenuis* (9) *C. ventricosa v. alpine V.* *(*Source*: Andhale 2009).**

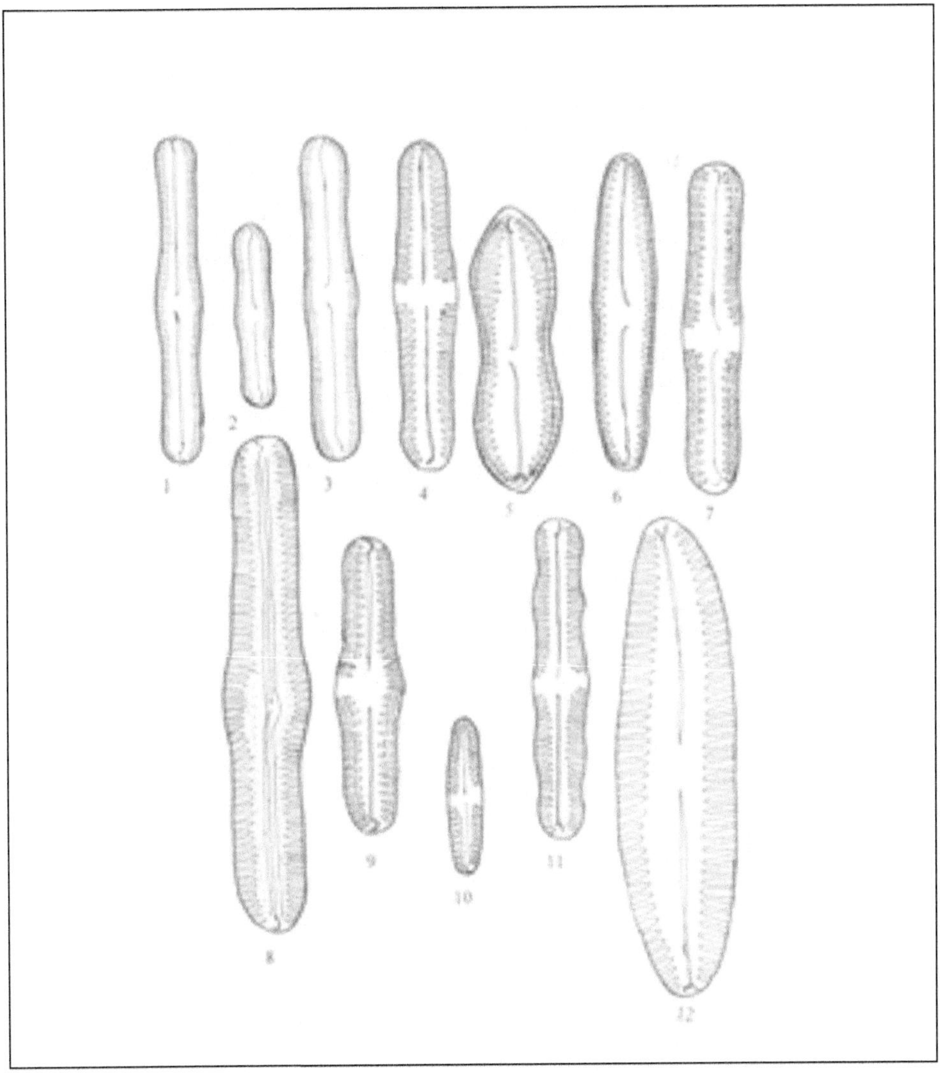

*Plate 33

Fig. (1) *Pinnularia acrosphaeria* (2) *P. acrosphaeria v. minor* (3) *P. acrosphaeria f. undulata* (4) *P. aestuarii v. interrupta* (5) *P. borealis v. lonavlensis* (6) *P.brevicostata v. indica* (7) *P. cardinaliculus* (8) *P. dolosa* (9) *P. kiusiuensis* (10) *P.kolhapurensis* (11) *P. legumen v. florentina* (12) *P. major v. linearis*
*(*Source*: Andhale 2008)

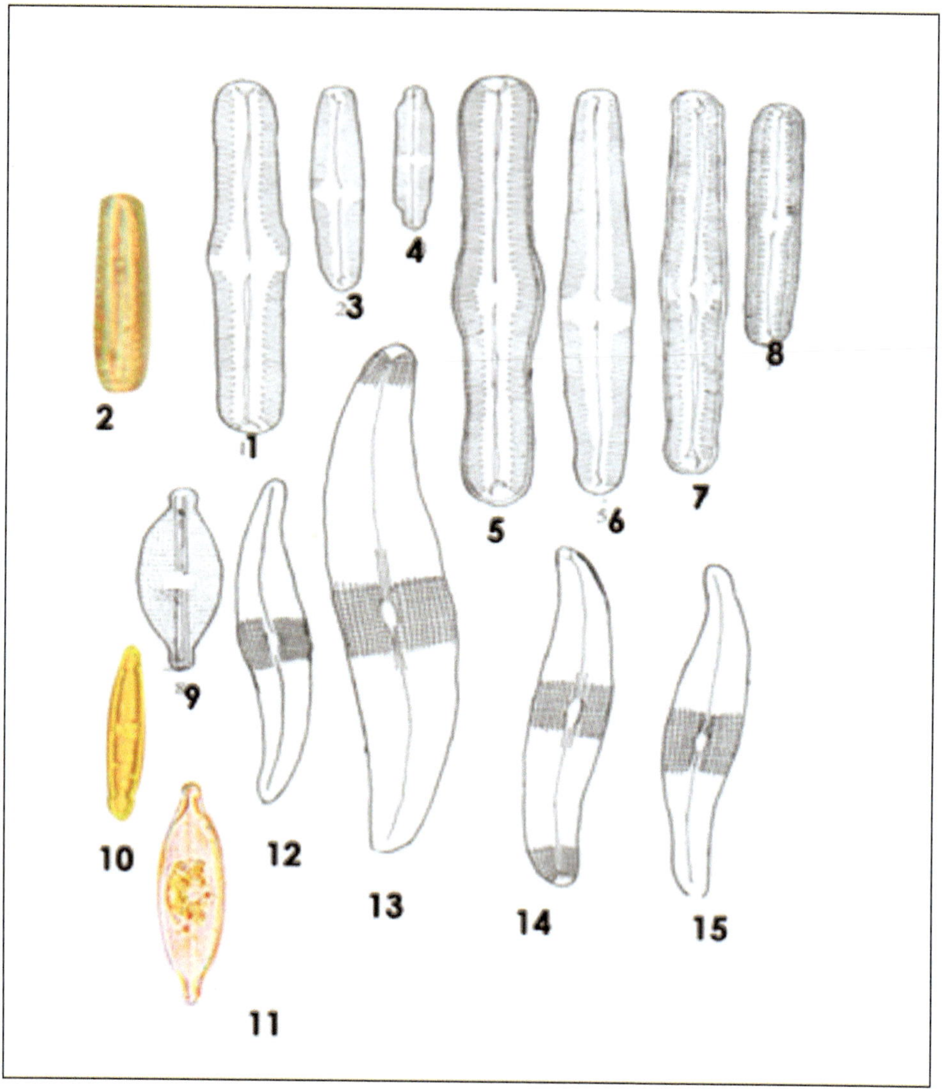


*Plate 34

Fig. (1) *P. marathwadensis* (2) *P. meisteri v. sandica* (3) *P. microstauron* (4) *P. notata v. rostrata* (5) *P. pseudoluculenta* (6) *P. stauropterav . Parva* (7) *P.stomatophoroides v. ornate f. erlangensis* (8) *P. sudetica v. commutate f. obtusata* (9) *Mastogloia baltica* (10) *M. recta* (11) *Frustulia jogensis* (12) *Gyrosigma acuminatum* (13) *G. bhusavalensis* (14) *G. khandeshensis* (15) *G. kuetzingii* *(*Source*: Andhale 2008).

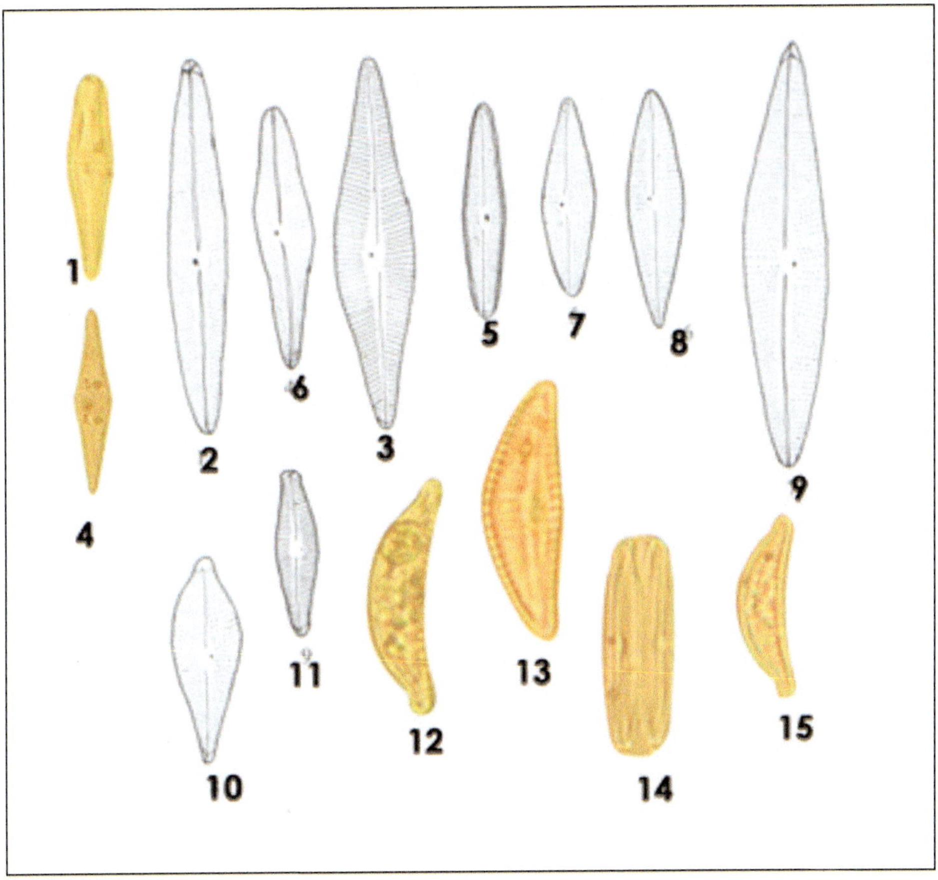


***Plate 35**

**Fig. (1) *Gomphonema aequatoriale* (2) *G. balatonis* (3) *G.gracile v. intricatiforme* (4) *G. gracile v. major* (5) *G. intricatum* 6) *G. lanceolatum* (7) *G. lanceolatum v. insignis* 8) *G. longiceps v. subclavata* (9) *G. magnifica* (10) *G. moniliforme* 11) *G.subventricosum* (12) *Amphora costata* (13) *A. maharashtrensis* (14) *A. ovalis* v. *gracilis* (15) *A. veneta* *(*Source*: Andhale 2008)**

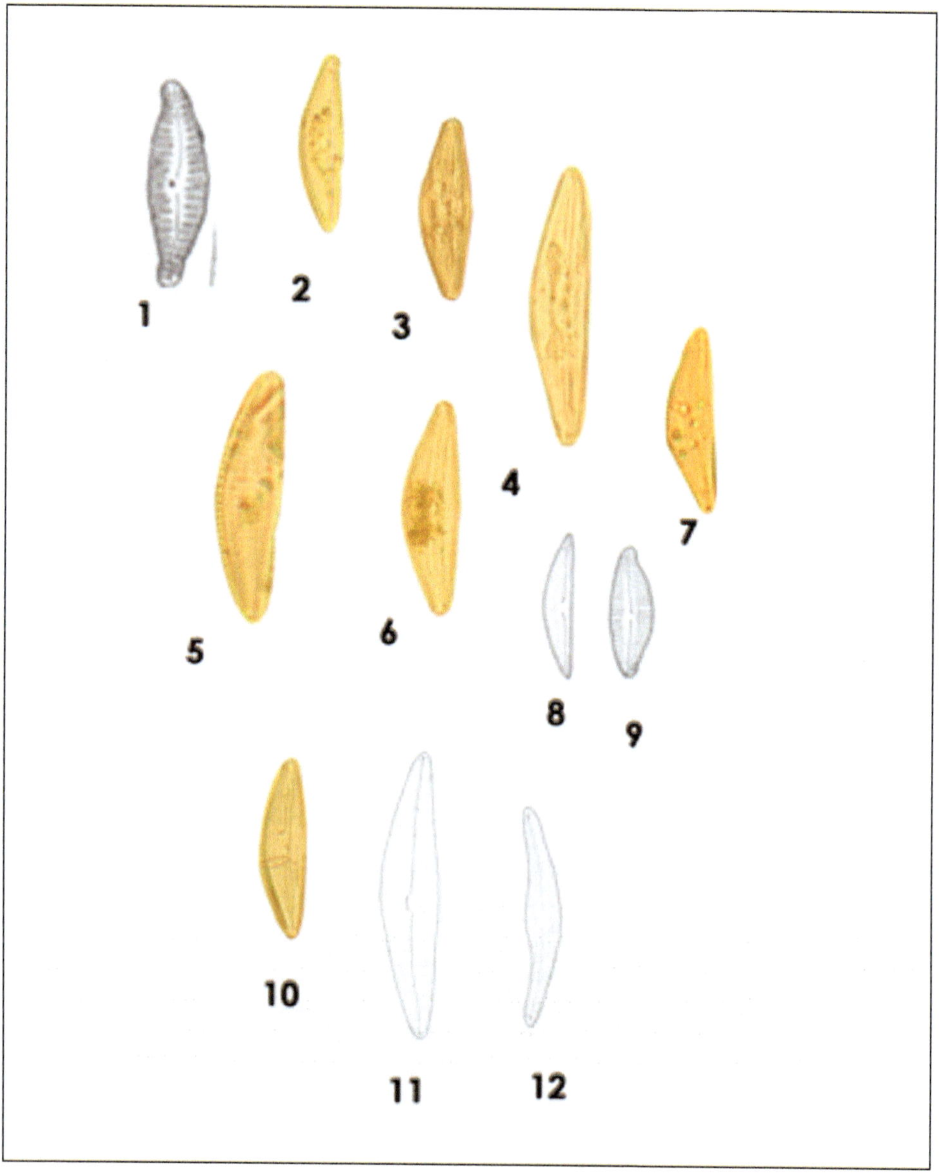


* Plate 36

Fig. (1) *Cymbella affinis* (2) *C. amphicephala* (3) *C. aspera* (4) *C. bengalensis* (5) *C.chandolensis* (6) *C. cistula v. woosungensis* (7) *C. gracilis* (8) *C. hungarica* v. *signata* (9) *C. kerkevarensis* (10) *C. laevis* (11) *C. lanceolata* (12) *C. lanceolata v. cornuta* *(*Source*: Andhale 2008)

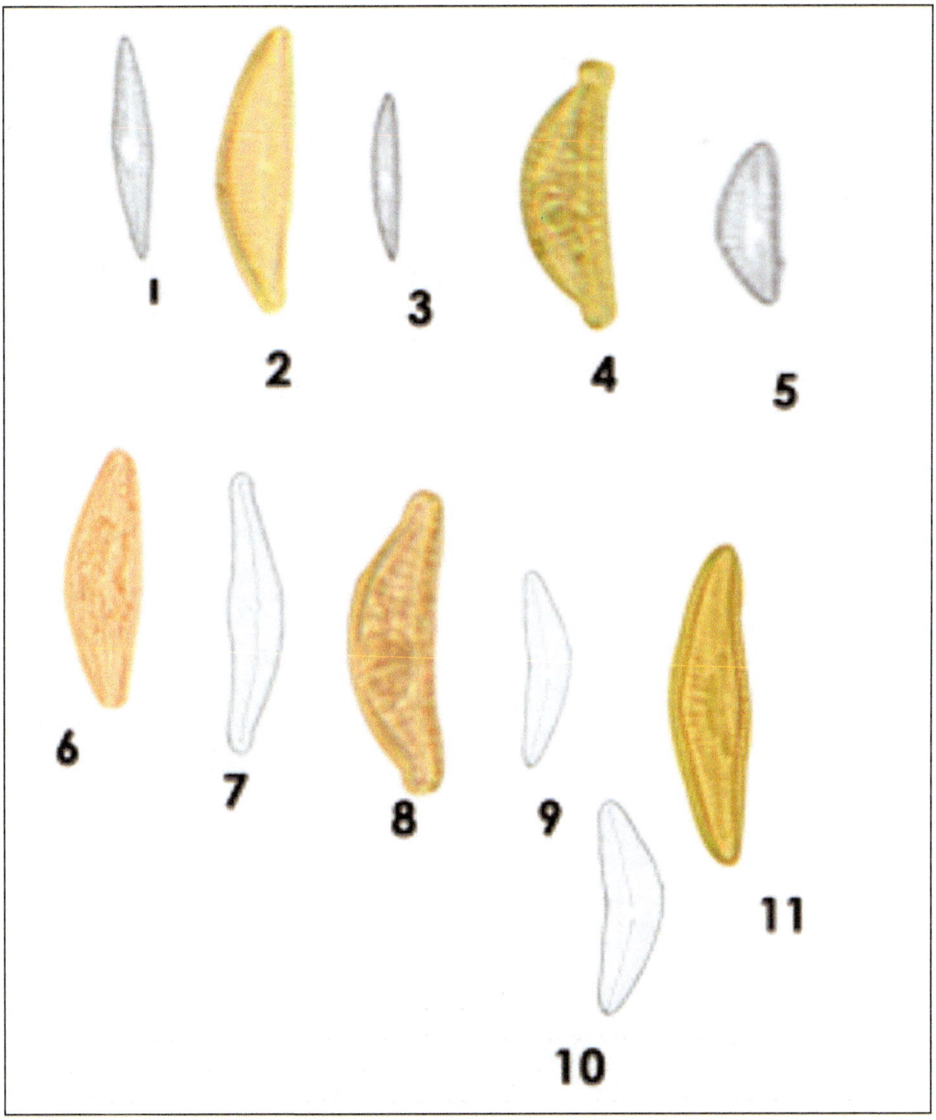


*Plate 37

Fig. (1) *C. maharashtrensis* (2) *C. osmanabadensis* (3) *C. perpusilla* (4) *C.powaiana* (5) *C.pusilla* (6) *C. tumida* (7) *C. tumida f. ventricose* (8) *C. turgidula* (9) *C. ventricasa* (10) *C. ventricosa v. depressa* (11) *C. vidarbhensis* *(*Source*: Andhale 2008)

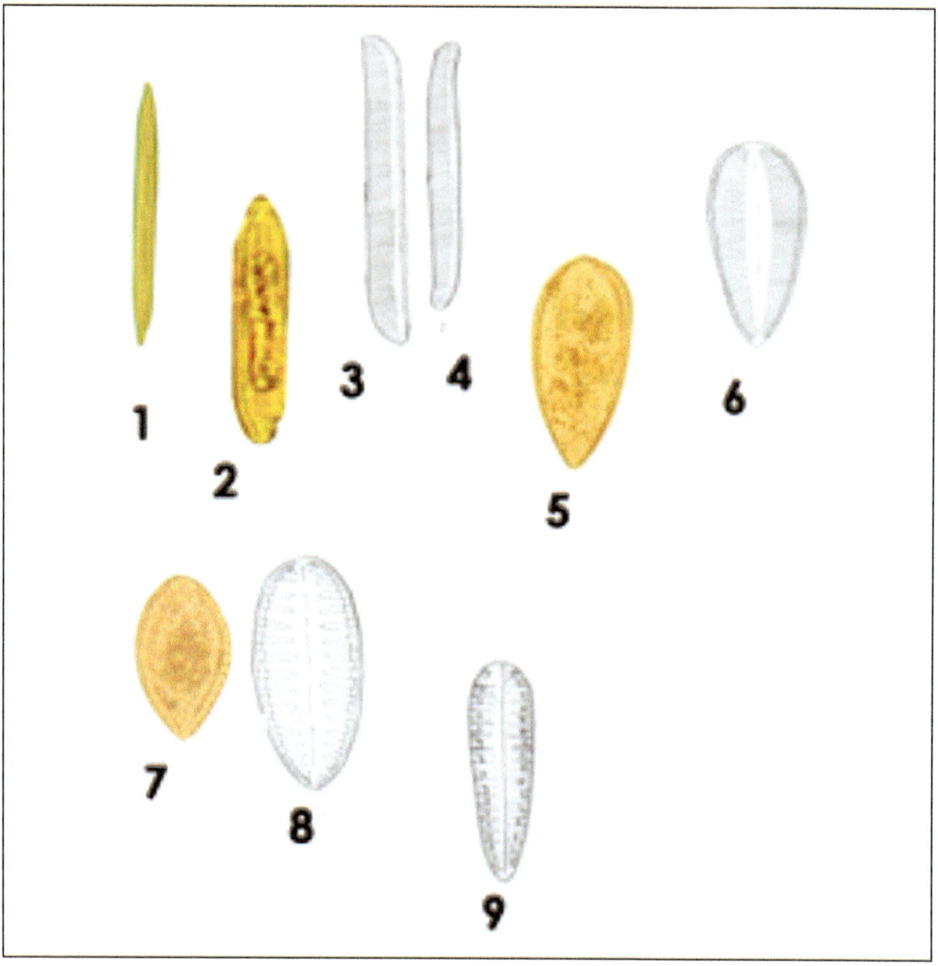


***Plate 38**

**Fig. (1)** ***Nitzschia intermedia*** **(2)** ***N. perminuta*** **(3)** ***N. obtuse*** **(4)** ***N. obtusa v. scalpelliformis*** **(5)** ***Surirella capronii*** **(6)** ***S. capronioides*** **(7)** ***S. ovata*** **(8)** ***S. robusta*** **(9)** ***S. tenuissima*** ***(*Source*: Andhale 2008)**

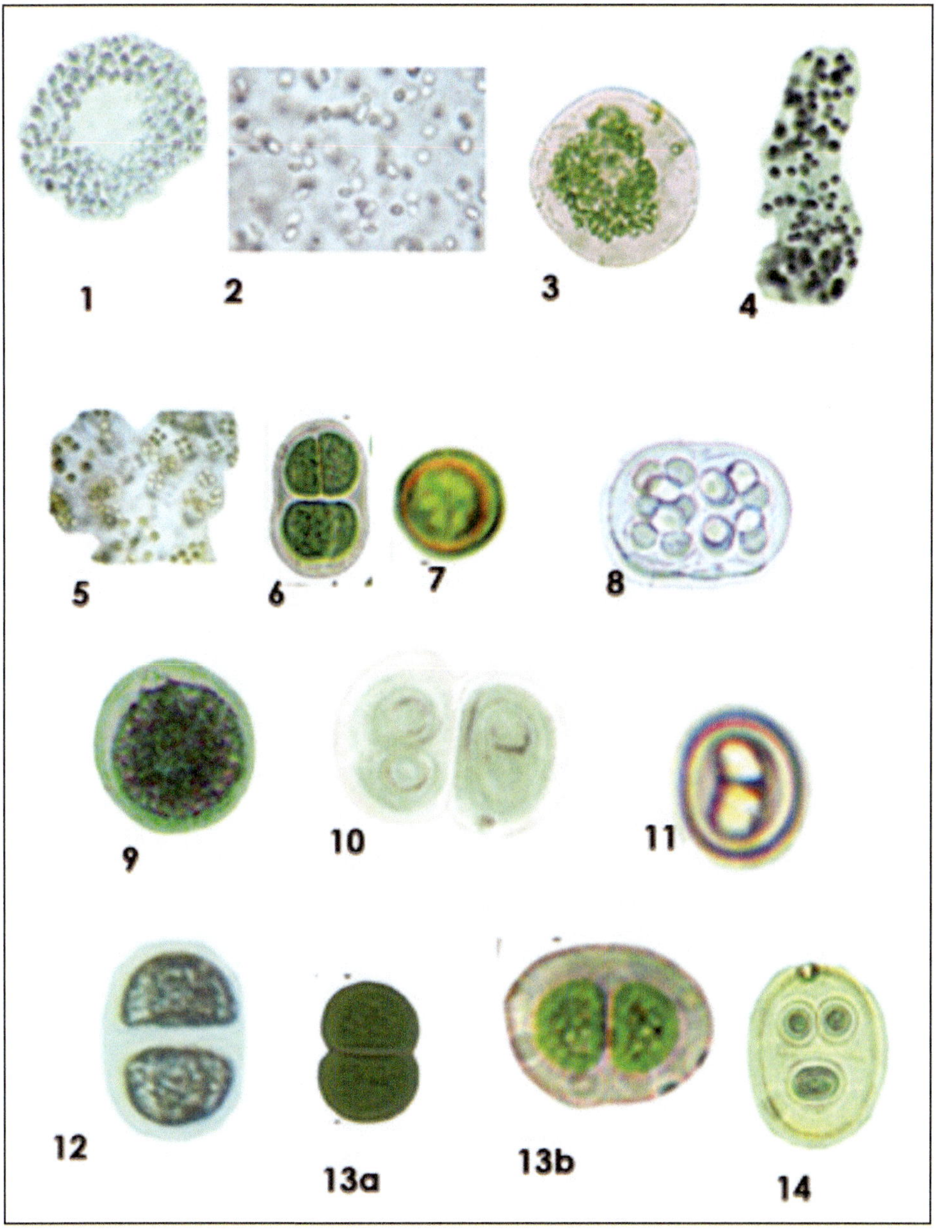


**Plate 39**

Fig. (1) *Microcystis aeruginosa* (2) *M. flos-aquae* (3) *M. marginata* (4) *M.pseudofilamentosa* (5) *M. robusta* (6) *Chroococcus giganteus* (7) *C. indicus* (8) *Ñ.limneticus* (9) *C. macrococcus* (10) *C. minutus* (11) *C. montanus* (12) *C.schizodermaticus* (13) *C. tenax* (14) *C. turgidus*

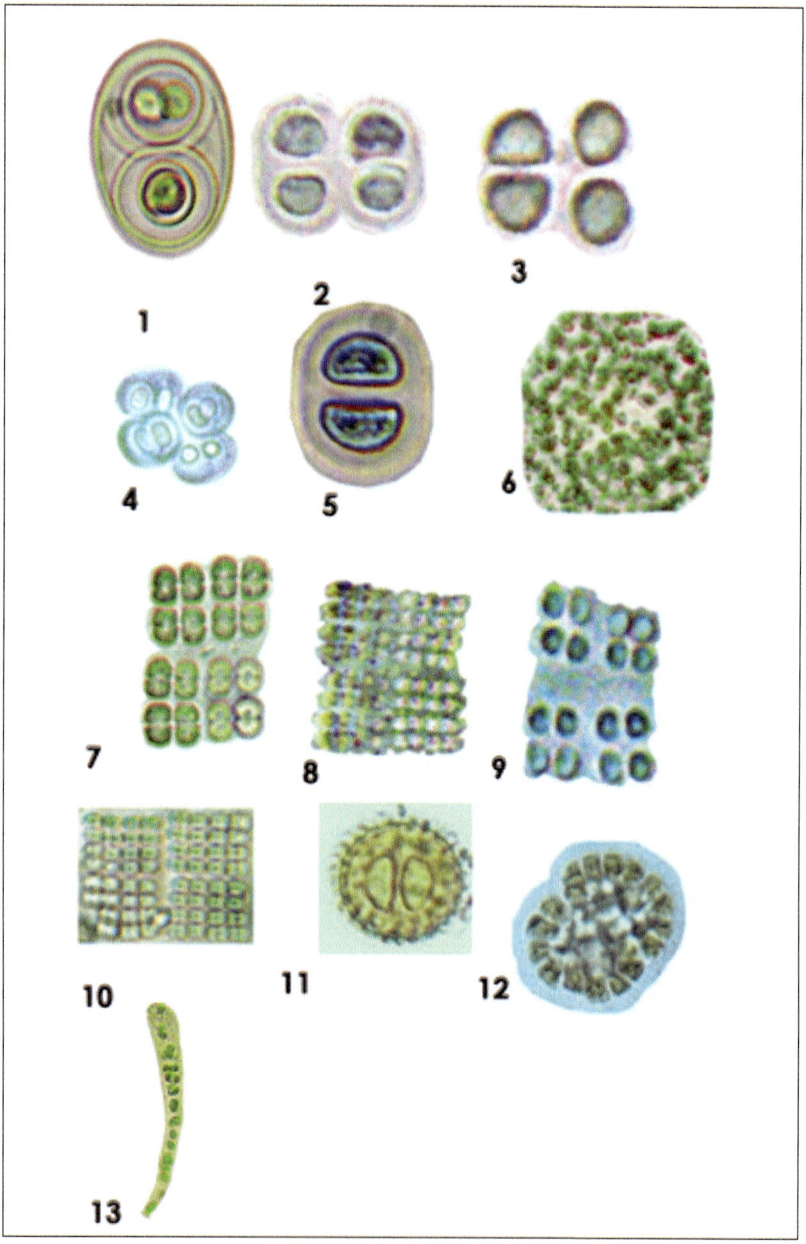


**Plate 40**

**Fig. (1)** ***Gloeocapsa calcarea*** **(2)** ***G. kuetzingiana*** **(3)*****G. magma*** **(4)** ***G. punctata*** **(5)** ***Aphanocapsa biformis*** **(6)** ***A. grevillei*** **(7)** ***Merismopedia elegans*** **(8)** ***M. glauca*** **(9)*****M. minima*** **(10)** ***M. punctata*** **(11)** ***Gloeothece rhodochlamys*** **(12)** ***Gomphosphaeriaaponina*** **(13)** ***Stichosiphon sansibaricus***

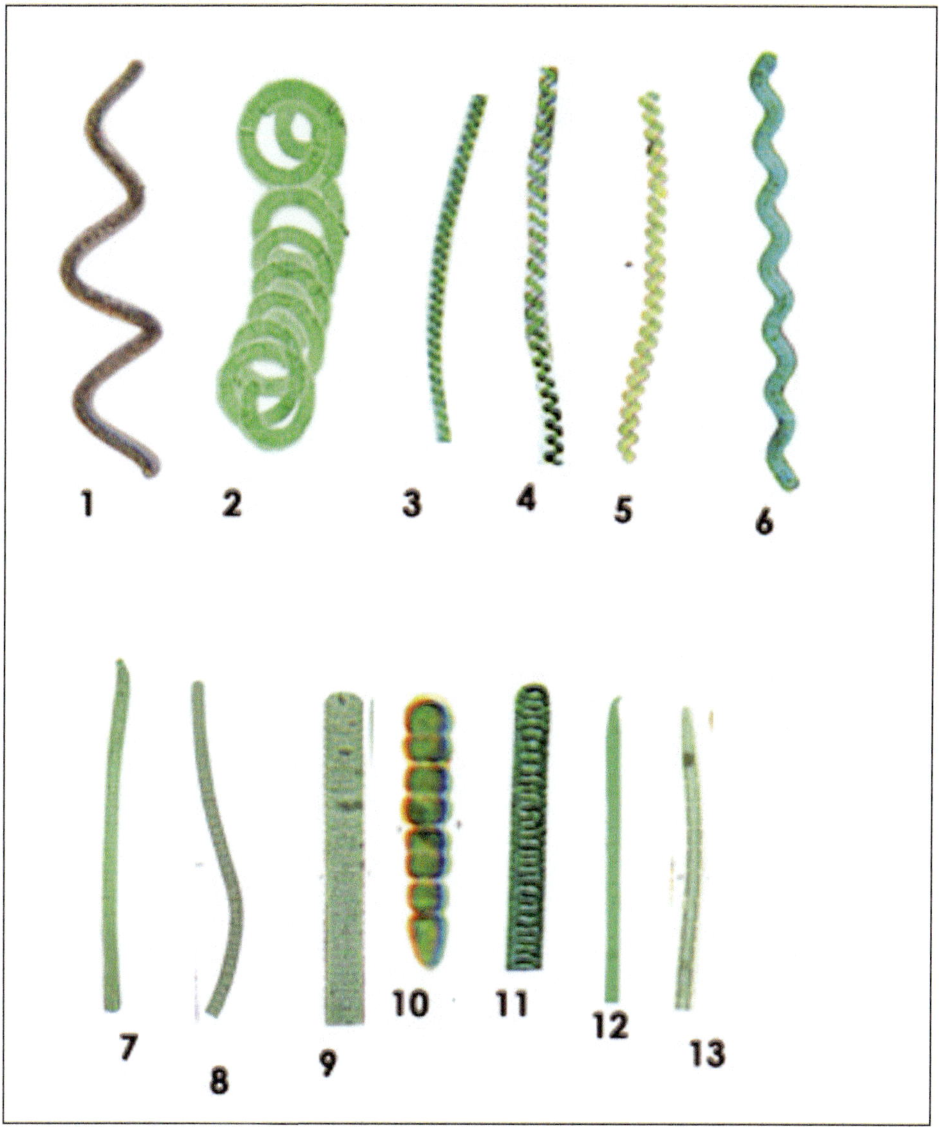

Plate 41

Fig. (1) *Arthrospira platensis f. granulata* (2) *A. platensis v. non- constricta* (3) *Spirulina gigantea* (4) *S. major* (5) *S. meneghiniana* (6) *S. princeps* (7) *Oscillatoria acuta* (8) *O. amphibia* (9) *O. annae* (10) 0. *biswasii* (11) *O. chlorina* (12) *O. cortiana* (13) 0. *cortiana v. minor*

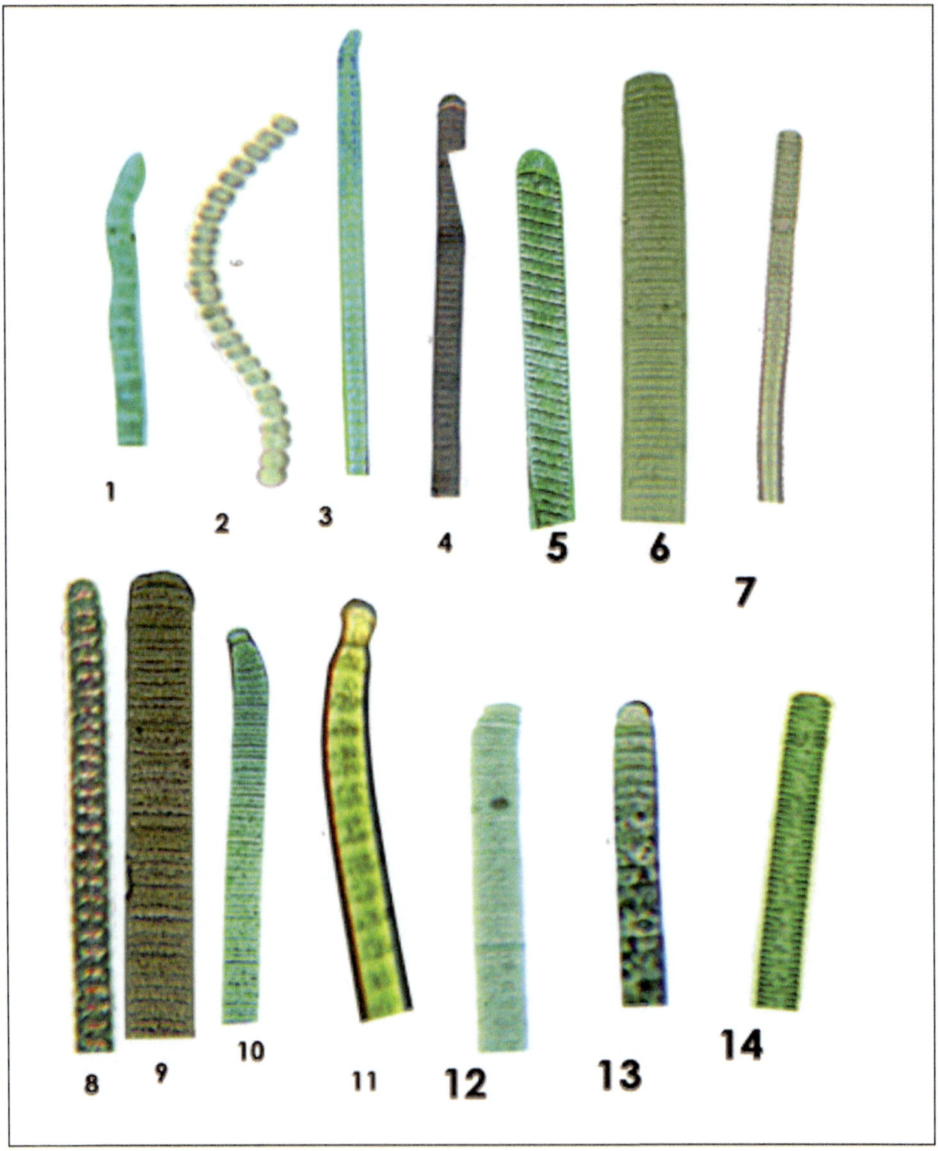


Plate 42

Fig. (1) *Oscillatoria curviceps* (2) *O. foreaui* (3) *O. formosa* (4) *O. irrigua* (5) *O. limosa*
(6) *O. margaritifera* (7) *O. ormata* (8) *O. ornata v. crassa* (9) *O. princeps*
(10) *O.proboscidea* (11) *O. rubescens V. curvata*
(12) *O.rubescens v. kolhapurensis* (13) *O. sancta* (14) *O. subbrevis*

**Plate 43**

Fig. (1) ***Lyngbya antartica v. maior*** (2) ***Lyngbya birgei*** (3) ***L. confervoides*** (4) ***L. connectens*** (5) ***L.hieronymusii*** (6) ***L. latissima*** (7) ***L.maharashtrensis*** (8) ***L. major*** (9) ***L. majuscula*** (10) ***L. martensiana***11) ***L. rubida*** (12) ***L. stagina f.non-granulata***

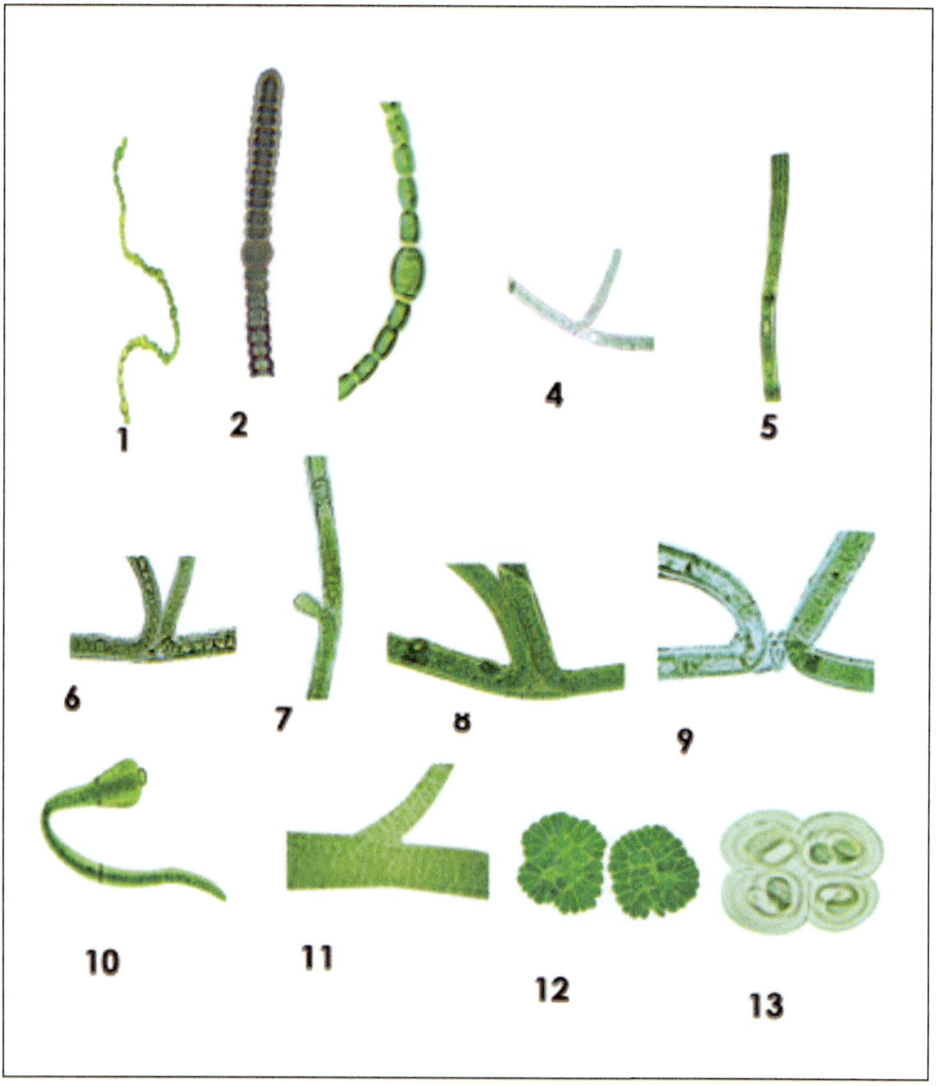


**Plate 44**

Fig. (1) *Cylindrospermum sphaerica* (2) *Anabaena iyengarii* (3) *Anabaena subcyl indrica* (4) *Scytonema arcangelii* (5) *S. bohneri* (6) *S. javanicum* (7) *S. ocellatum v. capitatum* (8) *S. simplex* (9) *Plectonema radiosum* (10) *Gloeotrichia raciborskii v. conica* (11) *Stigonema informe* (12) *Myxosarcina burmensis* (13) *M. spectabilis*

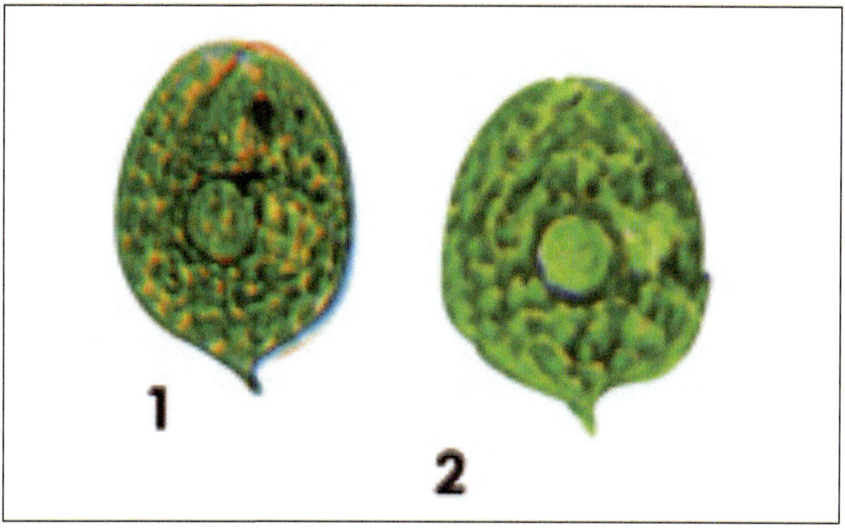


**Plate 45**
**Fig. (1)** ***Phacus curvicauda*** **(2)** ***P. pseudoswirenkoi***

# References


**Andhale, S.B.** (2008) Studies on the Flora of Jayakwadi Bird sanctuary, Ph. D. Thesis, Dr. Babasaheb Ambedkar MarathwadaUniversity, Aurangabad. **Andhale, S.B. and P.B.Papdiwal** (2006) *Azolla pinnata*R.Brown – A hitherto unreported water fern from Marathwada, *Bioinfolet***3** **(4)**: 323-324.

**Andhale, S.B. and P.B.Papdiwal** (2008) *Pediastrum* – A little star in Nathsagar water body, *Bioinfolet***5** **(3A):** 181-187.

**Andhale, S. B. and P.B.Papdiwal** (2009) Diversity of some Blue-green algae from Jayakwadi Bird Sanctuary, Paithan, Maharashtra. *J. Aqua. Biol.* 24 (**2**): 5-11.

**Andhale, S. B. and P.B.Papdiwal** (2009) Some Chlorococcalean algae from Jayakwadi Bird Sanctuary of Maharashtra. *J. Indian bot. Soc.***89** (1): 185-188.

**Ashtekar, P.V. (1980)** *Studies on the fresh water algae of*Chhatrapati Sambhaji Nagar district*. Ph.D. thesis, MarathwadaUniversity, Chhatrapati Sambhaji Nagar.

**Ashtekar, P.V. and N.D.Kamat (1978)** Additions to the desmidflora of Marathwada, Maharashta. *Phykos***18** (1 and 2): 45-50.

**Ashtekar, P.V. and N.D.Kamat (1979-a)** FilamentousMyxophyceae of Chhatrapati Sambhaji Nagar district,Maharashtra. *J. Bombay nat. Hist.Soc.* **76** (1): 215-218.

**Ashtekar, P.V. and N.D.Kamat (1979-b)** Croococcales ofChhatrapati Sambhaji Nagar district, Maharashtra. *MarathwadaUniversity Journal of Science,* **18** (11): 47-52.

**Ashtekar, P.V. and N.D.Kamat (1980)** Chlorococcales ofChhatrapati Sambhaji Nagar, Maharashtra.*Phykos***19** (1): 115-119.

**Deputy Conservator of Forests (Wildlife), ChhatrapatiSambhaji Nagar,** A publication entitled, "Jayakwadi BirdSanctuary".

**Desikachary, T.V. (1959)** *Cyanophyta*. Indian Council ofAgricultural Research, New Delhi.

**Dhande, J.S. and A.K. Jawale (2006)** On *Oedogonium* (Link)Hirn from Jalgaon District, Maharashtra. *Geobios***33**: 321-322.

**Falconer, I.R. (1999)** An overview of problems caused by toxicblue-green algae (Cyanobacteria) in drinking and recreationalwater. *Environmental Toxicology***14** (1): 5-12.

**Gonzalves, E.A. (1981)** *Oedogoniales,* Indian council of Agricultural Research, New Delhi.

**Jadhawar P.B. and P.B.Papdiwal (2011a)** Diversity of desmids at Nathsagar Water Reservoir *Bioinfolet***8** (3**)**: 280-284.

**Jadhavar P.B. and P. B. Papdiwal (2011b)** Diversity of aquatic angiosperms at Jayakwadi Bird Sanctuary water reservoir, *Life Sciences Bulletin,***8** (1**):** 53-56

**Jadhavar P. B. and P.B. Papdiwal (2011c)** Macro Algae at Nathsagar Reservoir, Paithan (M.S.) India, *Dn. Curr. Sci***6**: 384-386

**JadhavarP. B. and P.B. Papdiwal (2012a)** Taxonomic diversity of Diatoms at Nathsagar water reservoir, Paithan – Maharashtra. *The Ecoscan,* **6** (1 and 2): 1-4

**JadhavarP. B. and P.B. Papdiwal (2012b)** Some desmids from Nathsagar water reservoir, Paithan-Maharashtra. *J. Indian bot. Soc.* **91** (4): 317-321.

**Jawale,A.K. and J.S. Dhande (2005**) A Preliminary Survey of Chloroccocales From Hartala Lake-genus *Scenedesmus* Mayen. *Plant Diversity and Biotechnology* pp. 45-48.

**Jawale, A.K., D.A. Kumawat and N.A. Chaudhari (2010)**Some members of order Chlorococcales new toMaharashtra.*Bioinfolet***7** (2): 94-97.

**Kamat, N.D. (1963-a)** The algae of Kolhapur, India.*Hydrobiologia***22** (3-4): 209-305.

**Kamat, N.D. (1963-b)** The algae of Mahabaleshwar. *J. Uni. ofBombay***31** (3 and 5): 28-41.

**Kamat, N.D. (1963-c)** The Oscillatoriaceae of Ahmedabad, India.*J. Uni. of Bombay***31** (3 and 5): 20-27.

**Kamat, N.D. (1966-b)** Food of *Planispiranagporensis* Pfeiffer.*Curr. Sci.***35** (7): 179.

**Kamat, N.D. (1967-a)** The algae of Mount Abu, In, *"Theproceedings of the Rajasthan Academy of Sciences"***11** (1 and 2): 49-54.

**Kamat, N.D. (1967-b)** Dispersal of Charophytes by the Pintails,*Curr. Sci.* **36** (5): 134.

**Kamat, N.D. (1973-a)** Algae of Vidarbh, Maharashtra. *J. BombayNat. Hist. Soc.***72** (2): 450-476.

**Kamat, N.D. (1973-b)** Desmids of Marathwada, Maharashtra. *J.Bombay nat. Hist. Soc.***72:** 616-618.

**Kamat, N.D. (1974)** Algae of Marathwada, Maharashtra.*Phykos***13** (1)**:** 22-32

Kamat, N.D. (1982) *Topics in Algae,* Sai Kripa Prakashan,Chhatrapati Sambhaji Nagar.

**Korekar, S.L. and P.B. Papdiwal (2003)** A report on the algalflora from the forests of Osmanabad and Latur districts ofMaharashtra. In *"Biodiversity: Taxonomy and conservation",* (Eds M.A. Wadoodkhan, S.N. Jagtap. D. B. Gujrathi, D.P.Chavan,), Department of Botany Majalgaon College, Majalgaon,pp. 32-38.

**Kouwets, A.C. (1987)** Desmids from the Auvergne (France).*Hydrobiologia***146**: 193-263.

**Manickam, V.S. and V. Irudayarai (1962)** *Pteridophyte flora ofthe Western Ghats-South India.* B.I. Publications, New Delhi,pp.345.

**Misra, P.K., R.K. Mehrotra, Jai Prakashand A.K. S. Kishore (2001)** Fresh water green algae from Basti district, Uttar Pradesh.*Geophytology*31 (1 and 2): 1-7.

**Misra, P.K. and A.K. Srivastava (2003)** Some Desmids (Chlorophyceae) from North-Eastern Uttar Pradesh, India. *J.Indian bot. Soc.* **82**: 85-92.

**Misra, P.K., A.K. Srivastava, J. Prakash, D.K. Asthana andS.K. Rai (2005)** Some freshwater algae of Eastern Uttar Pradesh,India. *Our Nature*3: 77-80.

**Misra, P.K., R.K. Mehrotra, Jai Prakash, A.K. Srivastava andS. Kishore (2000)** Fresh water planktonic algae from Basti, UttarPradesh. *Geophytology***30** (1 and 2): 61-73.

**Misra, P.K., R.S.Chauhan and S.K.Shukla (2007)** Some fresh-water diatoms from Almora district, Uttaranchal State. *J.Indianbot. Soc.***86** (3 and 4): 74-79.

**Opute, F.I. (2000)** Contribution to the knowledge of algae ofNigeria. I. Desmids from the Warri/ForcadosEstuaries. Part II.The elongate baculiform desmids. *J. Limnal.* **59** (2): 131-155.

**Pal, B.P., B.C. Kundu, V.S. Sundaralingam and G.S.Venkataraman (1962)** *Charyophyta.* I.C.A.R., New Delhi.

**Pan, G., H. Chen, H. Zou, B. Tian, X. Yuan, J. Chen, M.Zhang and B. Yang. (2008)** Turning back catastrophic shifts inshallow lakes: Large scale Harmful Algal-Bloom (HAB) removaland submerged vegetation restoration using modified – local –soiltechnology. In: *"Proceedings of Taal 2007:* The 12th WorldLake Conference" (Eds. Sengupta, M and R. Dalwani) Jaipur, pp.906-913.

**Patel, R.J. and I. George (1977)** Chlorococcales of Gujarat,India- *Pediastrum* Meyen,*Sorastrum*Kuetzing and *Hydrodictyon*Roth. *J. Indian bot. Soc.* **56**: 172-178.

**Philipose, M.T. (1967)** *Chlorococcales,* Indian Council ofAgricultural Research, New Delhi.

**Prasad, B.N. and M.N. Srivastava (1992)** *Fresh water Algalflora of Andman and Nicobar Islands,* Vol-I, Bishen SinghMahendra Pal Singh, Dehra Dun.

**Prasad, B.N. and P.K. Misra (1992)** *Fresh water algal flora ofAndman and Nicobar Islands,* Vol. II, Bishen Singh Mahendra PalSingh, Dehra Dun.

**Prasad, B.N. and Poonam Khanna (1987)** The Cyanophyceaenflora of Sikkim-I. Oscillatoriaceae. *J. Indian bot. Soc.***66**: 253-260.

**Prasad, V.P., D. Mason, J.E. Marburger and C.R. AjithKumar (1996)** *Illustrated flora of Keoladeo National Park,Bharatpur, Rajasthan,* Bombay Natural History Society, OxfordUniversity Press, Mumbai.

**Prescott, G.W. (1951)** *Algae of the Western Ghat Lakes Area,*Cranbrook Institute of Science, Michigan.

**Prescott, G.W. (1964)** *How to know the fresh water algae* W.M.C. Brown Comapnay Publishers, Dubuque, Iowa.

**Purohit, S.S. (1990)** *Microbiology,*Agro Botanical Publishers (India), Bikaner, pp. 263.

**Rai, S.K. and P.K. Misra (2008)** On some desmids fromKoshi Tappu Wildlife Reserve, Nepal. *Ecoprint***15**: 47-58.

**Rai, S.K. R.K. Rai and N. Paudel (2008)** Desmids from Bees-hazaar Lake, Chitwan, Nepal. *Our Nature***6**: 58-66.

**Rajalakshmi, S., P.V. Deepa, A.K. Dixit and K. Kadavul (2007)** Seasonal distribution of phytoplankton in GautamiGodavari Estuary, Yaman, U.T. of Pondicherry. *The Bioscan***2** (4): 351-354.

**Ramnathan, K.R. (1964)** *Ulotrichales,* Indian Council ofAgricultural Research, New Delhi.

**Randhawa, M.S. (1959)** *Zygnemaceae,* Indian council ofAgricultural Research; New Delhi.

**Rao, R.S., P. Venkanna and T.A. Reddy (1986)** *Flora of Indiaseries, Flora of West Godavari district, Andhra Pradesh, India,*Indian Botanical Society, Meerut University, Meerut.

**Rath, J. and S.P. Adhikary, (2005)** *Algal flora of Chilika Lake,*Daya Publishing House, Delhi.

**Sahin, B. and B. Akar (2007)** The desmid flora of some highmountain lakes of the Turkish Eastern Black Sea region. *Pak. J.Bot.* **39** (5): 1817-1832.

**Salim Ali (2002)** *The Book of Indian Birds*, Bombay NaturalHistory Society, Oxford.

**Sarode, P.T. and N.D. Kamat (1984)** *Fresh water Diatoms ofMaharashtra*, SaikripaPrakashan, Chhatrapati Sambhaji Nagar.

**Scott, A.M. and G.W. Prescott (1961)** Indonesian Desmids.*Hydrobiologia*,17 (1-2): 1-132.

**Sen, N. and K. Naskar (2003)** *Algal flora of SundarabansMangals*, Daya Publishing House, Delhi.

**Smith, G.M. (1933, 1950)** *Fresh water algae of the UnitedStates*; McGraw-Hill Book Company, New York, London.

**Subramanyan, K. (1962)** *Aquatic Angiosperms*, Council ofScientific and Industrial Research, New Delhi.

**Suseela, M.R. and K. Toppo (2009)** Enumeration ofFresh wateralgal flora of Chandpur river of Palampur, Himachal Pradesh,India. *J. Econ. Taxon. Bot.* **33** (4): 966-972.

**Tiffany, L.H. and M.E. Britton (1952)** *The Algae of Illinois*,Hafner Publ. Co., New York.

*****West, W. and G.S. West (1905)** *A monograph of the BritishDesmidiaceae*, Ray Society, London, Vol.2, p 180.

# Index

## D



## E



## F

## G



## H



## K



## L



## M



## N

## O

## P



## R



## S